The Decisive Battles of World History

Gregory S. Aldrete, Ph.D.

PUBLISHED BY:

THE GREAT COURSES
Corporate Headquarters
4840 Westfields Boulevard, Suite 500
Chantilly, Virginia 20151-2299
Phone: 1-800-832-2412
Fax: 703-378-3819
www.thegreatcourses.com

Printed in the United States of America

Gregory S. Aldrete, Ph.D.

Frankenthal Professor of History and Humanistic Studies
University of Wisconsin–Green Bay

Professor Gregory S. Aldrete is the Frankenthal Professor of History and Humanistic Studies at the University of Wisconsin–Green Bay. He received his B.A. from Princeton University in 1988 and his Ph.D. from the University of Michigan in 1995. His interdisciplinary scholarship spans the fields of history, archaeology, art history, military history, and philology.

Among the books Professor Aldrete has written or edited are *Gestures and Acclamations in Ancient Rome*; *Floods of the Tiber in Ancient Rome*; *Daily Life in the Roman City: Rome, Pompeii, and Ostia*; *The Greenwood Encyclopedia of Daily Life: A Tour through History from Ancient Times to the Present*, volume 1, *The Ancient World*; *The Long Shadow of Antiquity: What Have the Greeks and Romans Done for Us?* (with Alicia Aldrete); and *Reconstructing Ancient Linen Body Armor: Unraveling the Linothorax Mystery* (with Scott Bartell and Alicia Aldrete).

Professor Aldrete has won many awards for his teaching, including two national ones: In 2012, he was named the Wisconsin Professor of the Year by the Council for Advancement and Support of Education (CASE) and the Carnegie Foundation for the Advancement of Teaching, and in 2010, he received the American Philological Association Award for Excellence in Teaching at the College Level (the national teaching award given annually by the professional association of Classics professors). Professor Aldrete also has been a University of Wisconsin System Teaching Fellow, a University of Wisconsin–Green Bay Teaching Scholar, and winner of a Teaching at Its Best award.

Professor Aldrete's research has been equally honored with a number of prestigious fellowships, including two year-long Humanities Fellowships from the National Endowment for the Humanities (NEH) and the Solmsen

Fellowship at the Institute for Research in the Humanities in Madison. Additionally, he was chosen as a fellow of two NEH seminars held at the American Academy in Rome; was a participant in an NEH institute at the University of California, Los Angeles; and was a Visiting Scholar at the American Academy in Rome. His university has given him its highest awards for both teaching and research: the Faculty Award for Excellence in Teaching and the Faculty Award for Excellence in Scholarship, both from the Founders Association.

Professor Aldrete's innovative Linothorax Project, in which he and his students reconstructed and tested ancient linen body armor, has recently garnered considerable attention from the media, having been featured in documentaries on the Discovery Channel and the Smithsonian Channel and on television programs in Canada and across Europe. It also has been the subject of articles in *U.S. News & World Report*, *Der Spiegel*, and *Military History* and of Internet news stories in more than two dozen countries.

Professor Aldrete maintains an active lecture schedule, including speaking to retirement groups; in elementary, middle, and high schools; and on cruise ships. He also has been named a national lecturer for the Archaeological Institute of America. For The Great Courses, he taught *History of the Ancient World: A Global Perspective*. ■

Table of Contents

INTRODUCTION

LECTURE GUIDES

Table of Contents

Table of Contents

Table of Contents

SUPPLEMENTAL MATERIAL

The Decisive Battles of World History

Scope:

Many of the most decisive turning points in the history of the world have been battles. More than just conflicts between armies, such moments often represent fundamental clashes among rival religions; cultures; and social, political, and economic systems. The outcomes of these battles have dramatically transformed and shaped the course of history, often sending it on unexpected or completely new paths. This course examines more than three dozen such pivotal moments, highlighting and exposing the key incidents and personalities responsible for these critical shifts. During the course of these lectures, we'll discover how the Battle of Yarmouk contributed to the establishment of Islam in the Middle East; how the Battle of the Talas River curbed the expansion of the Tang dynasty of China; how the Battle of Boyaca resulted in South American independence from Spain; and how the Battle of Khalkhin Gol, fought on the borders of Mongolia and Manchuria, influenced the entire direction of World War II in both Europe and the Pacific.

This course features three aspects that should be relatively original, even for those with some familiarity with military history. First, it is truly global in scope, including not only the more familiar battles of Western civilization but also pivotal ones in Asia, South America, India, and the Middle East. Thus, we'll cover Mohamad of Ghor and the Battle of Tarain in India, as well as William the Conqueror and the Battle of Hastings; we'll explore the Battle of Sacheon in Korea, as well as the Battle of Stalingrad. Second, the course analyzes both key land battles and naval clashes, topics that are frequently treated separately. Third, although many famous battles are included, there are also a good number that are not very well known, such as the battles of Cynocephalae, Yarmouk, Diu, and Ayacucho. Often, a more obscure battle whose outcome was actually more decisive is substituted for a much better known but, in reality, less pivotal one. Thus, instead of Marathon, we'll look at Plataea; rather than Waterloo, we'll explore Leipzig; and in place of Gettysburg, we'll discuss Antietam.

Naturally, the lectures provide clear and vivid accounts of the campaigns and battles themselves, but they also offer in-depth descriptions of the cultural aspects of warfare, including the nature of the societies involved. We'll come to understand, for example, how the code of samurai behavior shaped the outcome of the Battle of Sekigahara or how the attitudes of the Crusaders contributed to their defeat at the Battle of Hattin.

Similarly, most lectures feature an examination of the often colorful personalities who played crucial roles in the conflicts, whether generals, politicians, soldiers, or inventors. We will witness, on the one hand, how the impetuosity of young Ramesses II brought him victory, while for the Prussian von Moltke, it was his coldly calculating mind that led to success. The biographies of these key individuals are filled with memorable incidents and interesting trivia. Thus, we'll see how Horatio Nelson's brilliant naval career nearly came to a premature end in the jaws of a polar bear and how the entire course of the war in the Pacific during World War II might have been completely different if, in his youth, Admiral Yamamoto had lost three fingers during a battle rather than two.

This course reveals the secrets behind of some of the most famous armies of all time, such as those of Alexander the Great and Genghis Khan, explaining the tactics and technologies that allowed them to triumph over their foes. It also traces the effects of changing technologies over time and shows how an edge in technology frequently resulted in military success, from the hoplite style of warfare of the ancient Greeks, to the innovative turtle ships of the Korean Admiral Yi, to the steel swords and primitive muskets of the Spanish conquistadors. Although perhaps unfortunate, it is nevertheless true that warfare typically sparks technological creativity and invention—consider the advance of the airplane from flimsy fabric and wood biplanes to jets in only five years during World War II. Similarly, the most sophisticated products of technology often are found in the military. In examining the great battles of human history, we will also trace the overall history of technological innovation, from the Stone Age to the dawn of the Space Age.

Those who enjoy military history will find much to appreciate in this course, but it will also be of great interest to anyone with a basic desire to understand why history turned out as it did and how we got to where we are today. Battles have served as the catalysts for many of the key turning points in the human story, and it is impossible to fully comprehend the development of civilizations, religions, technology, and cultural movements without considering the place of warfare in determining the course of events. Spanning the entire globe and all eras, this engaging series of lectures reveals the profound and eternal impact of decisive battles in human affairs. ■

What Makes a Battle Decisive?

Lecture 1

Many forces can influence the course of history—great ideas, economic trends, demographic shifts—but one of the most frequent and dramatic is warfare. One obvious explanation for the widespread existence of war throughout human history is its potential for causing rapid change, and within warfare, the most concentrated form of change is individual battles. It is this potential to rapidly alter the status quo and initiate dramatic shifts in fortune or dominance that causes battles to be identified as turning points in history. This is the idea behind this course: to examine some of the key battles that, for one reason or another, have signaled fundamental shifts in the direction of events.

When Is a "Decisive Battle" Not?

- On July 20, 1866, just off the coast of modern Croatia, Admiral Wilhelm von Tegetthoff led an Austrian naval squadron against an Italian fleet that was both technologically and numerically superior. Undeterred, Tegetthoff arranged his ships into an arrowhead formation and boldly drove them straight at the long, menacing line of Italian warships. This confrontation, which would be known as the Battle of Lissa, had all the makings of one of the decisive battles in history:
    - Its immediate outcome would determine the fate of the city of Venice.
    - It would decide who would control the Mediterranean Sea.
    - It was part of a larger confrontation between two grand coalitions of nations.
    - It marked the first time that a major sea battle was fought between large numbers of ironclads, a potent new form of naval vessel that promised to instantly render all previous wooden warships obsolete.

- The turning point of the battle came when Tegetthoff used his flagship to ram one of the most powerful Italian ironclads, the *Affondatore*. The attack ripped a huge hole in the side of the Italian vessel and, within minutes, the stricken ship rolled over and went to the bottom.

- Yet today, the Battle of Lissa is hardly remembered. Why? First, its strategic importance as an Austrian victory was eclipsed when, in the same month, the Battle of Könnigrätz delivered a crushing defeat at the hands of the Prussians, resulting in the fall of the Austrian Empire. Second, the conclusions that naval strategists drew from Lissa were completely wrong.
  - The battle was interpreted as establishing the dominance of ramming as a tactic in future naval warfare, with the effect that all major warships for the next 40 years were built with rams. In reality, the utility of ramming was an anomaly.

  - For several generations, huge battleships still sported ludicrous and useless rams, even though naval battles of the next half century would be fought at increasingly long ranges by massive cannons.

Features of Decisive Battles

- In this course, we will examine famous and not-so-famous battles, generals, tactics, strategies, weapons, and wars. Yet it is also a course about historical causation: why things turned out the way they did and how sometimes the most significant events turned on, or were determined by, the very smallest of acts or chances.

- If we look at the entire span of human warfare, twists of fate at pivotal moments turn out to be common. Consider:
  - In the 17th century, did the theft of an officer's horse, which caused him to fail to make his customary nightly patrol, allow a successful surprise attack, with the effect that most of North America became British territory rather than a colony of France?

- During the American Civil War, did a messenger's carelessness result in the loss of vital battle plans, contributing to the ultimate defeat of the Confederacy?

- During World War II, did a faulty mechanism on an aircraft-launching catapult cause a fatal half-hour delay in launching a single plane, resulting in the destruction of the Japanese navy and overall American victory in the Pacific?

- In certain instances, we will explore slightly more obscure battles rather than more famous ones if an interesting case can be made for the decisiveness of the less well-known battle. For example, from the Napoleonic wars, we will look at Leipzig rather than Waterloo.

- What makes a battle decisive?
  - First, it was one that was militarily decisive in that the defeat of one military force by another resulted in an immediate and obvious transfer of political power. A variant of this type is a decisive battle that results in the near or total destruction of a vital component of an opponent's forces. Major naval battles, with their concentration of high-value units in one place, are especially prone to fall into this category. The Battle of Trafalgar, for example, had a profound effect on the rest of Napoleon's career: The loss of his fleet definitively crushed his plans to invade England and drove him to the fatal decision to invade Russia instead.

  - Second, perhaps the most common type of decisive battle is one that subsequently had important social, political, or religious effects. In many cases, these battles may not have seemed pivotal at the time but have been recognized only in retrospect as demarcating a turning point. For example, the American War of Independence would have ended much sooner but for Washington's daring crossing of the Delaware River and success at the Battle of Trenton. If not for this unlikely victory, the young American Republic would have been snuffed out of existence before it ever really got going.

Other Considerations

- Over the next 36 lectures, we will rummage through nearly 4,000 years of history and travel all around the globe looking for key turning points. During our search, we will examine both land battles and naval clashes, and we will consider some battles that involved millions of participants, while others featured just a handful of people. Some of our battles were immediately recognized at the time as being important transitional moments, while for others, their true importance was acknowledged only much later.

- The list tends to favor battles that curbed or ended the growth of various expansionist empires because without such key defeats, those empires might well have extended their political and cultural domination yet further. The siege of Vienna in 1683, for example, represents the high-water mark of expansion for the Ottoman Empire.

- Another consideration in favor was for a battle whose outcome was either unexpected or uncertain. We will look at battles in which the sides were roughly evenly matched or for which it is easily possible to imagine a dramatically different outcome.

- Finally, some battles were selected as decisive because they represent the introduction of a key technological advance or the triumph of one type of military force over another. In the technology category could be considered the Battle of Midway, which set the pattern for future naval clashes being decided by air power rather than big guns. Of the second type, the Battle of Cynoscephalae revealed the superiority of the Roman military system over the previously dominant Hellenistic one and, thus, heralded Rome's ascension over, and conquest of, the entire Mediterranean basin.

- Many of the battles we will look at could be placed into more than one of these categories. Also, strictly speaking, some of the battles presented here could be considered campaigns, and occasionally, we will lump together several closely related battles that resulted in a collective outcome.

© Willard/iStock/Thinkstock.

From the invention of the wheel through the development of the jet engine, typically, the first uses of new inventions are in military contexts.

- Although this course is by no means a comprehensive history of warfare, it will inevitably trace or, perhaps more accurately, mirror the development of weapons, strategy, and tactics over time.

- A recurring theme of the lectures is technological change. The mighty steel dreadnaughts of the First World War—direct descendants of the *Affondatore*—boasted cannons that could fling tons of explosive shells 20 miles in one broadside. These ships were so expensive that the naval arms race between Great Britain and Germany nearly bankrupted both countries. When we examine the Battle of Midway, we will encounter aircraft carriers, which represented another key technological shift. Each of these types of vessel represented the most cutting-edge technology and was among the most expensive mobile manmade objects of its day.

- Examining the course of history by focusing on the idea of finding decisive battles can be a useful analytical tool because it encourages us to view history not as a boring and immutable timeline but, instead, as a series of constantly branching pathways whose outcomes and effects are frequently unpredictable and whose real significance often emerges only with the passage of time.

Suggested Reading

Creasy, *Fifteen Decisive Battles of the World.*

Davis, *100 Decisive Battles from Ancient Times to the Present.*

Holmes, *Battlefield.*

Weir, *50 Battles That Changed the World.*

Questions to Consider

1. What qualities or characteristics do you think determine whether or not a battle deserves to be called "decisive"?

2. Do you agree that individual battles can truly change the course of history, or are these events just symptomatic of broader underlying forces?

1274 B.C. Kadesh—Greatest Chariot Battle

Lecture 2

In late May of 1274 B.C., on the banks of the river Orontes in Syria, the young ruler of Egypt, Ramesses II, rode at the head of a vast Egyptian army and was on the verge of leading them into a battle against his archenemy, King Muwatalli of the Hittites. Ramesses confidently anticipated winning a victory that would catapult him into the ranks of the greatest among Egypt's long line of glorious pharaohs. Unknowingly, however, the eager pharaoh was riding into a trap. The resultant clash would become known as the Battle of Kadesh, and it is the earliest battle in human history whose course and maneuvers we can reconstruct in detail.

Background to Kadesh

- In the generations leading up to the Battle of Kadesh, the main challenger to Egypt for supremacy in the eastern Mediterranean was the powerful Hittite empire based in Anatolia, a region that roughly corresponds to modern Turkey.

- Lying between was Syria, a strategic and economic crossroads that connected the Mediterranean basin to Mesopotamia and, therefore, became a hotly contested territory desired by both the Hittites and the Egyptians and regarded by both as lying within their zone of influence.

- A minor kingdom in this region usually controlled the key fortified city of Kadesh and played the Hittites and the Egyptians against each other.

- From a broader historical perspective, the Battle of Kadesh is notable not only for being our earliest detailed battle account but for several other significant characteristics:
    - It was one of the largest chariot battles in history.

- It resulted in one of the world's earliest peace treaties whose full terms have survived.

- It affected the course of ancient Near Eastern history for centuries.

- It formed the cornerstone of the reputation of one of Egypt's most famous pharaohs.

The Opponents

- Ramesses II was 29 years old and had ascended to the throne of Egypt during the period known as the New Kingdom, when Egypt became an imperialist power that sought to extend its sphere of influence south along the upper Nile into Africa and east and north along the Mediterranean coast.

- Ramesses's father, Seti I, pursued an aggressively expansionist policy, personally leading several large military expeditions, crushing a coalition of Canaanite princes to conquer Palestine, and pushing north into Lebanon. Yet Kadesh had slipped from his grasp into alliance with the Hittites.

- Ramesses was eager to establish his own reputation as a great military pharaoh, and he already showed signs of being a charismatic leader and a competent strategist. For example, as soon as he gained the throne, he began to build up the army and establish forward bases that would provide essential logistical support for any major campaigns.

- Muwatalli, Ramesses's Hittite opponent, was not originally intended to inherit the Hittite throne; he became king after his older brother died. Muwatalli's father had fought against Seti; thus, the ongoing conflict between Egypt and the Hittites had something of a generational aspect.

The Egyptian Army and Its Technology

- The Egyptian army consisted of four divisions of infantry, each composed of 5,000 men and several thousand chariots. Modern estimates suggest that perhaps 500 chariots accompanied each of the divisions and constituted the elite strike force.

- The three components necessary for an effective war chariot are the spoked wheel, horses, and a reasonably powerful bow. These technologies seem to have come together in the 2nd millennium B.C. and spread via nomadic Indo-European tribes throughout Eurasia.

- By the time of Ramesses, Egyptian chariots were highly refined war machines.
    - They emphasized speed and mobility and favored a lightweight design with six spoked wheels with narrow rims, a D-shaped cab made of ox hide stretched over a light wooden frame with rawhide strips for a floor, and a single long pole to which the horses were yoked.

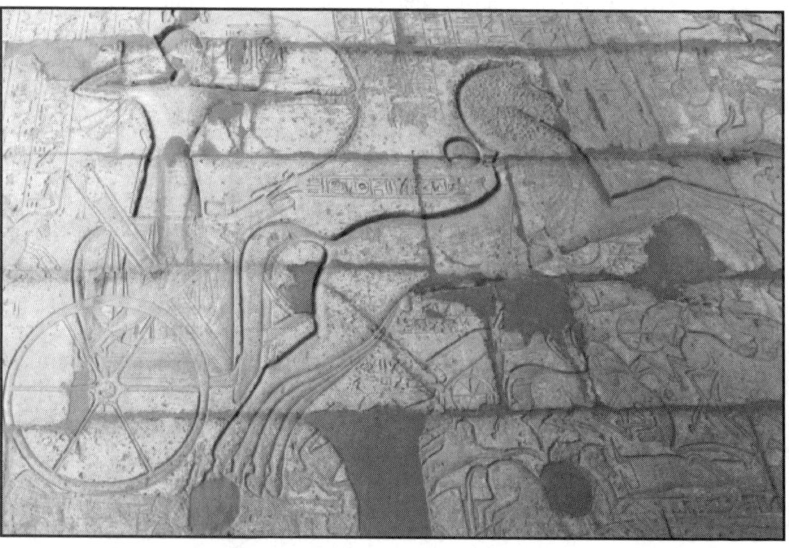

Egyptian chariots are frequently and vividly depicted in Egyptian art, giving us a good idea of their appearance and construction.

- The axle was set far to the rear for stability and a smaller turning radius, and the crew consisted of two men: a driver who also carried a shield and a warrior armed with a compound bow and javelins.

- These vehicles could be readily broken down and carried by infantry in order to traverse rough ground. Their main purpose was to serve as rapidly moving archery platforms that could charge, spin around, retreat, and charge again while unleashing flights of deadly arrows.

- In addition to being the higher-status units of the army, the charioteers were also some of the more highly trained soldiers. When he accompanied the army, the pharaoh naturally assumed the role of the lead charioteer.

- Although there was a permanent professional core to the army, in times of war, the majority of the ranks, especially the infantry, were filled out with temporary recruits, often farmers. They were armed with simple spears, bows, or a bronze axe or sword. Body armor was minimal, perhaps a skullcap or jerkin made of stiffened fabric or leather.

The Hittite Army and Its Technology

- Muwatalli had assembled a gigantic army, with an estimated size of 30,000 to 40,000 men and several thousand chariots. One source claims that there were 3,500 of these; if that number is accurate, this battle may well have been the largest clash of chariots in history.

- The standard Hittite chariot was significantly different from the Egyptian version:
  - It had a much heavier design, with a larger and more solid rectangular wooden cab.

  - The axle was centered beneath the cab rather than to the rear.

- It carried three crewmen: the driver, a warrior with a large shield and spear, and one wielding a long thrusting spear or a bow.

- In battle, the Hittites favored a single mass charge by the heavy chariots, intended to break the enemy's ranks and then run them down by using the spears or shooting arrows.

The Battle

- As Ramesses and the Egyptian army approached Kadesh, they were unaware that the Hittite army was nearby; thus, for ease of marching, the four divisions were spaced at intervals of roughly a half day's march. Ramesses accompanied the lead division, Ammon. Behind them came the Ra, the P'tah, and the Set divisions.

- When Ammon crossed the Orontes, two men who appeared to be Bedouin locals but were actually Hittite agents told Ramesses that Muwatalli had been frightened at the Egyptians' approach and had fled to the north. Ramesses accepted their story and evidently made no attempt to confirm the information with his own scouts.

- The camp guards then caught two Hittite spies lurking nearby who, after being subjected to a thorough beating, divulged that Muwatalli was not fleeing in terror but was hiding just on the other side of Kadesh and that his army was ready for battle.

- Ramesses immediately dispatched messengers with orders for all the elements of his scattered forces to converge on his location at maximum speed.

- Muwatalli sent forward a strong contingent of his chariots to intercept the Ra division as it attempted to march to the rescue, catching the Egyptians strung out in marching formation. The heavy Hittite chariots swept through the protective screen of lighter Egyptian chariots, slammed into the lines of marching infantry, and carved a path through the center of the Egyptian formation. The surviving troops panicked, broke formation, and ran.

- Following the retreating remnants of the Ra division, the Hittite chariots charged the Ammon camp, overrunning the shield wall formed to defend the camp. The charge took them into the midst of a maze of tents and wagons, piles of supplies, and military gear.

- The chaos bought Ramesses vital time to arm and organize himself and gather the chariots of the Ammon division, supplemented by the surviving Ra chariots. He then led the assembled Egyptian chariots in a counterattack against the now distracted and disorganized Hittite chariots.

- Seeing the fight turning against him, Muwatalli ordered his personal entourage of chariots into the fray. Their effect was negated by the timely and long-awaited arrival of the Egyptian reinforcements. The first to arrive was a chariot force known as the Ne'arin, perhaps accompanied by advance units of the third Egyptian division. Assaulted from a new direction by the Ne'arin, the Hittites broke and fled.

- The arrival of the main body of the P'tah division late in the day and the reorganization of the surviving Ammon and Ra units further tilted the balance in Ramesses's favor and effectively ended the battle.

- Although the vast majority of his army, including all the infantry, had not been engaged, Muwatalli withdrew within the walls of Kadesh. Ramesses had seemingly snatched a battlefield victory from the jaws of defeat.

- The day after the battle, Ramesses signed a truce with Muwatalli and returned with his army to Egypt. The Hittites retained control of Kadesh.

Who Won the Battle of Kadesh?

- In a narrow tactical sense, Ramesses can be considered the victor on the battlefield, but in a broader strategic sense, the Hittites won the campaign because they accomplished the main goal of the war: possession of the city of Kadesh.

- Although this battle was less decisive in military terms than others, it permanently ended the multigenerational war between two of the greatest powers of the era: Sixteen years after the battle, Ramesses and the Hittites signed a remarkable peace treaty.
    - It contains provisions establishing borders, mutual declarations not to invade each other's territory, promises of support if one or the other country is attacked or to help suppress internal rebellions, and extradition of political refugees.
    - The treaty ushered in an unprecedented era of peace in the ancient Near East that would last nearly a century.
- The Battle of Kadesh is also important because it served as the foundation of Ramesses's reputation as a leader, which he would amply exploit over the course of a 66-year reign. During this time, he built many of the most famous monuments of ancient Egypt.
- In the next lecture, we will move north along the shores of the Mediterranean to examine a pivotal battle fought between the ancient Greeks and the mighty empire of Persia.

Suggested Reading

Cotterell, *Chariot.*

Gardiner, *The Kadesh Inscriptions of Ramesses II.*

Goedicke, ed., *Perspectives on the Battle of Kadesh.*

Healy, *The Warrior Pharaoh.*

Shaw and Boatright, "Ancient Egyptian Warfare."

Questions to Consider

1. What do you think are the pros and cons of chariot warfare?

2. In what ways did the personality of Ramesses affect the battle, taking into account his actions before and during it?

479 B.C. Plataea—Greece Wins Freedom

Lecture 3

Some battles are decisive because of what they prevent from happening. The Battle of Plataea, which took place in 479 B.C. and was fought between the united city-states of ancient Greece and the Persian Empire, is one of these. If the Greeks had lost this battle and become merely one more province of the Persian Empire, the cultural flourishing of Greece in the 5th century B.C. might not have taken place. At the very least, a Persian victory would have resulted in a different course of history.

Background to Plataea

- Plataea is not nearly so well-known as three other battles fought between the Greeks and Persians within an 11-year span. Thermopylae was a Greek defeat, and Marathon and Salamis, although Greek victories, were only temporary setbacks for Persia, which returned to the fight each time.

- Plataea, however, was decisive. It effectively ended the war and ensured Greek independence and freedom, thus making possible the Greek golden age.

The Opponents

- On the one side was mighty Persia, a culturally sophisticated, ethnically diverse, and economically prosperous empire that stretched from the Mediterranean to the borders of modern India.

- Pitted against this colossus were the Greek city-states, a group of small, separate political entities on the mainland of Greece and the islands of the Aegean Sea that shared a common language and culture.

- The largest was Athens, known for its boldness and creativity, which had begun to experiment with forms of democracy. Next was Sparta, inward-looking, suspicious, and possessed of a small but terrifyingly efficient and fanatical army. These two spent most of their time engaged in fierce squabbles with each other.

The Greek Army and Its Technology

- During the 6th century B.C., a military innovation occurred in Greece: the hoplite revolution, a style of fighting in which heavily armed and armored foot soldiers fought in an organized formation, carrying a heavy, circular, concave shield three feet in diameter and shaped so the hoplites could nestle their shoulders and torsos within the curve.

© Bibi Saint-Pol/Wikimedia Commons/Public Domain.

Hoplite warfare worked best when each man acted as an identical and interchangeable cog in the war machine of the phalanx.

- The other standard piece of equipment was a long stabbing spear equipped with a bronze spearhead and a smaller bronze butt-spike that could function both as an alternate spear point if the main one broke off and for downward thrusts.

- The hoplite was a formidable opponent, protected from the front from head to toe in solid armor. Yet the weight of his shield and armor made him cumbersome, and he was vulnerable to attack from the sides and behind.

- The solution to this vulnerability was the phalanx: long rows, several men deep, with their shields close to one another or at times even overlapping. Fighting as a phalanx, each man in essence protected his neighbor, and as long as the phalanx kept its cohesion and no one allowed a gap to open, it was highly effective.

- Some historians believe that the hoplite revolution led to a so-called "Western way of war" that emphasized well-organized heavy infantry and decisive battles with the purpose of killing one's foes and that this style of warfare accounts for much of the success of Western Europe in conquering the globe during the era of colonialism.

The Persian Army and Its Technology

- The Persian army reflected the ethnic diversity of the Persian Empire, including a wide range of troop types, weapons, and armor. The bulk of the army comprised temporary conscripts, but it also included a number of more professional contingents. The Persians also employed mercenaries in their army, including large contingents of Greek hoplites.

- Among the professional soldiers were the Immortals, a well-trained, cohesive group of 10,000 elite infantry. Their shields, often made of wicker or animal hides, offered far less protection compared to the heavy wood and bronze carried by the typical hoplite, and their body armor was similarly lightweight.

- Some of the best units in the Persian army were the cavalry, considered the most prestigious arm of the military and, thus, appropriate for the aristocracy. Persian horsemen wielded light spears, axes, and swords, but their armor was relatively light.

- The principal weapon of the Persians was the bow, used by both foot and mounted archers, and Persian military tactics often featured harassing squadrons of horse archers who would charge in, release flights of arrows, retreat, and then circle back for another charge.

Precipitating Events

- Conflict between Persia and Greece arose with the revolt of some eastern Greek cities that had been absorbed by the Persian Empire. Although the insurrection was crushed, the rebels had received some aid from the Greek mainland. The Persian king viewed this aid as unwarranted interference and launched a punitive expedition in 490 B.C.

- At the subsequent Battle of Marathon, the Persians suffered a surprise defeat at the hands of the Athenians. Although Marathon was important for stopping this first Persian invasion and for demonstrating the superiority of the hoplite style of warfare in hand-to-hand combat, from the Persian perspective, it was a minor setback for a small expeditionary force.

- The Persians recognized that they would have to send another army to crush the Greeks, but internal politics delayed their return for 10 years. In 480 B.C., a massive land and sea invasion force crossed the Hellespont, led personally by Xerxes, king of Persia, intending to overwhelm and subjugate the Greeks.

- The Greeks attempted to stop the Persians by occupying the pass of Thermopylae, whose narrow confines nullified their superior numbers as a factor. It was a good strategy, undone when a traitor showed the Persians an alternate route through the mountains. Although a rear guard of 300 Spartans volunteered to stay and hold off the Persians while the others escaped, they were slaughtered, fighting to the last man.

- The Battle of Thermopylae vividly demonstrated the bravery of the Spartans, but it did nothing to stop the Persian advance, which reached central Greece. The Athenians were forced to flee their city, and the Persians occupied and burned it.

- The Greeks recycled their Thermopylae strategy at sea, opposing the Persian navy in a narrow strait between the mainland and the island of Salamis. The resulting naval battle was a success for the Greeks, and a significant portion of the Persian fleet was destroyed. Yet the victory of Salamis did not end the invasion and left the real threat to Greece, the vast Persian land army, untouched and still occupying central Greece.

- What Salamis did accomplish was to complicate supplying the huge Persian army. Accordingly, Xerxes decided to return to Persia with many of the conscripts, while leaving behind the best elements of his army to complete the conquest of southern Greece. The general in charge of this task was Mardonius, an experienced military commander.

The Battle

- Mardonius chose the 10,000 Immortals, as well as large infantry and cavalry contingents, as his army. The resulting force was, in many ways, more dangerous than the bloated force that had invaded Greece, and it was still much larger than any army the Greeks could collectively muster.

- Mardonius camped for the winter, during which he made several attempts to break up the Greek alliance by exploiting traditional rivalries and suspicions. This strategy almost worked, but most of the Greeks united and marched north against him.

- Mardonius took up a position along the Asopos River, near Thebes, which had thrown in its lot with the Persians out of jealousy of Athens. The best modern guess is that probably 80,000 to100,000 Greeks squared off against about 100,000 to 150,000 Persians, Thebans, and other pro-Persian Greeks.

- Mardonius made the first move, sending some elements of his strong cavalry forces to harass the Greeks and search for a weak spot. During this clash, the popular leader of the Persian cavalry, a man named Masistos, was unhorsed and killed when an arrow killed his mount. After some bitter skirmishing over the body, the Persians retreated, leaving the trophy in the hands of the Greeks.

- After a week or so of standoff, Mardonius began sending elements of his cavalry on raids behind the Greek line, harassing their supply trains and eventually capturing one of their key water sources. Running short on food and water, the Greek commanders decided to pull back during the night to a well-watered and more defensible area called “the island.”

- But the Greeks bungled the retreat. One contingent of refused to obey the orders to leave; thus, once the Greeks finally began to move, they were strung out.

- Seeing an opportunity to destroy them, Mardonius ordered a general advance, and the Persians and their allies swept forward. Almost unintentionally, the main battle was joined.

- The battle came down to a savage close-quarters shoving match in which desperate Persians grabbed and broke the Greeks' spears. In this melee, the heavier armor of the Greeks gave them an advantage. Mardonius and his bodyguard were killed, along with many of the best of the Persian troops. The battle turned into a rout, with the triumphant Greeks chasing and slaughtering the defeated Persians.

Outcomes

- On the same day as the Battle of Plataea, a naval battle against the remnant of the Persian fleet also gave victory to the Greeks, and this moment marked the end of the Persian threat. The war would continue for decades, but it was the Greeks who were on the offensive.

- The victory at Plataea ushered in a period known as the Pentekonteia, a 50-year period regarded as the golden age of Greece that ended when they once more fell prey to their old rivalries, resulting in the disastrous 30-year Peloponnesian War.

- To commemorate Plataea, the Greeks melted down some of the Persian weapons and used the bronze to erect a column at Delphi. It was stolen 800 years later by Constantine and moved to Constantinople, where it decorated the horse-racing arena. It can still be seen in Istanbul, and it still legibly bears the names of the 31 Greek cities that united to fight at the Battle of Plataea.

Suggested Reading

Connolly, *Greece and Rome at War*.

De Souza, Heckel, and Llewellyn-Jones, *The Greeks at War*.

Green, *The Greco-Persian Wars*.

Herodotus, *The Landmark Herodotus*.

———, *Histories Book IX*.

Shepherd, *Plataea 479 BC*.

Questions to Consider

1. In what ways might history have been different if Persia had won the Battle of Plataea and conquered Greece?

2. What are the strengths and weaknesses of the hoplite style of warfare?

331 B.C. Gaugamela—Alexander's Genius

Lecture 4

What are the qualities that make an outstanding general? Intelligence? Creativity? Daring? Calculation? Charisma? Luck? The key figure at Gaugamela amply possessed all of these and is often regarded as one of the greatest generals of all time. He conquered most of the known world, was victorious in four major battles, conducted several successful sieges, and held together a multinational army during an epic march across much of Europe and Asia. His name, of course, was Alexander, more commonly known as Alexander the Great.

Background to Gaugamela

- The Battle of Gaugamela, perhaps Alexander's greatest victory, demonstrates that one of the keys to his success as a general was his unusual combination of cautious preparation before battle and quick-thinking boldness once engaged.

- The golden age of the Greeks had ended with the long, destructive Peloponnesian War, in which the Greeks again turned against one another. Although Sparta was the nominal victor, all participants were exhausted and impoverished by the struggle.

- Over the next half century, a new power arose to the north: Macedonia, a minor, disunited state, weakly controlled by a hereditary king. Between 359 B.C., when Philip II came to the throne, and 339 B.C., he transformed Macedonia into a

By all accounts, Alexander combined a sharp intelligence with great athleticism; he was a charismatic figure with the gift of inspiring intense loyalty among his followers.

first-rate power. Most notably, he reconstructed the army, which he then used to conquer his neighbors and create a Macedonian empire.

- When Philip was assassinated in 336 B.C., Alexander succeeded to the throne at the age of 20. The young king quickly secured his position, getting the army to swear an oath of loyalty to him personally, killing any who might be potential rivals, and suppressing several revolts.

The Opponents

- At the time of his death, Philip had been planning an expedition against Persia's westernmost regions, ostensibly in revenge for Persia's invasions of Greece more than 100 years earlier. Alexander now took up this plan, and in the spring of 334 B.C., he crossed the Hellespont into Asia at the head of a Macedonian-Greek army of approximately 45,000 infantry and 5,000 cavalry.

- The Persian Empire at the time of Alexander unquestionably remained the superpower of the region. The current king of kings was Darius III, who had come to the throne in 336 B.C. The ancient sources (all of which, admittedly, are Greek) offer conflicting portraits of Darius, with most depicting him as weak and indecisive. However, in the first year of his reign, he successfully put down a rebellion in Egypt.

The Macedonian Army and Its Technology

- In revamping the Macedonian army, Philip made a number of innovations to the successful model of the Greek hoplite phalanx:
    - He equipped the Macedonian phalanx with an extra-long spear called a sarissa, rather than the shorter spear of the hoplite.

    - He lightened the soldiers' armor, particularly reducing the size and weight of the shield.

- Maintaining order in wielding these weapons required considerable drilling and discipline; thus, Philip made his army a permanent, professional one. He also included contingents of other types of troops:

- He added archers and slingers to harass the enemy from a distance. There were swift, agile, lightly armed troops to act as skirmishers.

- There were sizable cavalry units, some of which were light cavalry used as scouts, while others were heavy cavalry who could break an enemy line.

- Such a force, known as a mixed army, had two significant effects.
  - It made the new Macedonian army much more flexible, able to fight against a range of enemies and to react to a variety of circumstances and conditions.

  - It put greater emphasis on good generalship. A clever general might give separate missions to different parts of his army or send units in different directions. Used creatively, it was an army with great potential.

The Persian Army and Its Technology

- The Persian army was much the same as that which the Greeks had faced at Plataea. Its elite infantry were the 10,000 Immortals.

- The strength of the Persian army was its numerous and well-trained cavalry, and a favored weapon among both infantry and horsemen was the bow, although a wide array of swords, spears, and axes were also employed.

Precipitating Events

- After crossing into Asia and visiting Troy, Alexander began his invasion of the Persian Empire. The governor of the region organized the local forces, including a number of Greek hoplite mercenaries, and marched out to confront him. At the Battle of the Granicus, Alexander led a charge and won the victory. He then proceeded deep into Asia Minor, conquering cities as he went.

- Recognizing Alexander as a serious threat, Darius determined to take the field himself and gathered a large army to intercept the Macedonians. They met in 333 B.C. at the Battle of Issus, and although outnumbered, Alexander again prevailed, personally leading the charge.

- Rather than immediately pursuing Darius, Alexander spent several years subduing the Persian fleet and capturing all of its bases. Darius spent this time preparing for their next confrontation. Both knew that the next time they met would be the decisive battle to determine which was to rule the Persian Empire.

The Battle

- Darius's most obvious advantage was numerical, and he used the interval to gather a vast army from every corner of his empire.

- Darius then chose his battlefield carefully: a large, flat, featureless expanse along the Tigris River near Gaugamela, where no topographical features could anchor Alexander's flanks and where the Persian numerical superiority could be used to full effect.

- Finally, Darius created a special weapon of 200 chariots with blades attached to their wheels. These would be launched against the Macedonian phalanx; the blades would literally carve openings in the formations, into which the cavalry could pour.

- Alexander's advisors, frightened by the size of the Persian army, urged him to attack at night to mask their inferior numbers. Refusing this advice, Alexander went to bed. Darius, fearing just such an assault, kept his army standing ready for battle all night. By morning, Alexander had already scored an advantage: His troops were well rested in contrast to the sleepless Persians.

- Alexander's most pressing problem was the lack of a geographic anchor to prevent encirclement.

- His solution was to stagger his forces at an angle on the left and right sides of the phalanx so that they could face an enemy encircling movement head on.

- At a distance behind the main phalanx, he positioned a second line made up of allies and mercenaries.

- His forces were arranged so that if the Persians did outflank them and surround his army, it could form a hollow rectangle with men facing outward in all directions.

- As the battle began, Alexander led his cavalry to the right. Darius ordered his cavalry to mirror Alexander's movements, with the result that the lines were stretched out and the center of gravity began to shift away from the ground that Darius had so carefully prepared.

- Darius therefore ordered his scythed chariots to charge, but the Macedonian skirmishers and javeliners picked off the charioteers. When the remainder reached the phalanx, the Macedonians opened lanes for the chariots to pass harmlessly. As they slowed to turn, lightly armed troops killed the rest of the charioteers.

- Determined to contain Alexander's sweep, Darius dispatched more Persian cavalry to block him, and an intense cavalry battle ensued. Meanwhile, on the left, the Macedonian phalanx was hard pressed and in danger of losing contact with Alexander and the Macedonian right.

- A gap developed in the Macedonian line, into which a group of Persian cavalry poured. Had they wheeled to the right and struck the main phalanx from behind, they might well have broken the phalanx and won the battle. But the Persian cavalry began looting the Macedonian baggage train; thus, elements of the second line of the Macedonian phalanx had time to confront and contain them.

- Meanwhile, Alexander had decided to take advantage of the parallel stretching of the Persian lines, and he now led a bold charge, cutting back from the right flank toward the center of the Persian formation, where Darius stood in his chariot. The Companion cavalry, formed into a wedge with Alexander himself at its head, crashed into the Persian ranks.

- Darius apparently took fright and, just as he had at the earlier Battle of Issus, turned his chariot and fled from the field, abandoning his army. Upon his desertion, the Persians in that section of the field lost heart, and after some tough fighting, Alexander and his cavalry routed them.

- Alexander had won the battle and, with it, the Persian Empire. Although Darius escaped and would manage to evade capture for another year or so, until mortally wounded by his own men, from the moment that he fled Gaugamela, he had effectively forfeited his throne; Alexander became the new king of kings of Persia.

Outcomes

- Gaugamela was Alexander's finest achievement as a general. It was a battle in which his enemy seemed to have significant advantages: superiority in numbers, choice of battlefield, and a secret weapon. But through a combination of carefully preparing and training his army and skillfully and boldly using them during the battle, Alexander nullified these Persian advantages one by one.

- Alexander is sometimes accused of impulsiveness, but he showed calculation after Issus in not pursuing Darius and, instead, neutralizing the Persian navy to secure his flanks and protect his supply lines. Even in the heat of combat at Gaugamela, he demonstrated sound judgment in curtailing his charge and going to the aid of his phalanx.

- After Gaugamela, Alexander went on to capture the royal cities of Persia. Even though his united empire did not survive beyond his own lifetime, his real legacy was the spread of Greek culture throughout the empire. Indeed, the next period of Mediterranean history is termed the Hellenistic era because of the domination of Greek, or Hellenic, culture.

- If not for Alexander's victory at Gaugamela, Greek civilization might never have spread beyond the boundaries of Greece and, thus, would not have exerted its pivotal influence on the course of Western civilization.

Suggested Reading

Engels, *Alexander the Great and the Logistics of the Macedonian Army.*

Fox, *Alexander the Great.*

Griffith, "Alexander's Generalship at Gaugamela."

Heckel and Jones, *Macedonian Warrior.*

Sekunda and Warry, *Alexander the Great.*

Questions to Consider

1. How much of Alexander's success do you think was the result of Philip's actions?

2. How much of a role did Alexander's personality play in his success?

197 B.C. Cynoscephalae—Legion vs. Phalanx

Lecture 5

Other than a scattering of Greek colonies, the poorer, less civilized western Mediterranean had not been much of a concern for the Hellenistic kingdoms of the 3rd century B.C., but by its end, a new power had arisen in the west. These were the Romans, who had slowly spread from their city beside the Tiber to conquer most of Italy. They had then fought two bitter wars against their rival, Carthage, to emerge as the dominant force in the western Mediterranean. Now these upstarts were showing an inclination to spread into the eastern part of the sea.

Background to Cynoscephalae

- In the autumn of 197 B.C., two armies representing two very different military systems engaged in a battle that would determine which of those systems was superior, with far-reaching consequences for world history. One of these armies was led by King Philip V of Macedon. The second was a Roman one commanded by Titus Quinctius Flamininus.

- The Romans had enjoyed success against their mostly barbarian foes in the west, but now they were encountering the highly skilled, professional, and experienced armies of the east. Because the two forces were almost exactly the same size, the battle constituted an important showdown between rival military styles and systems, with nothing less than the domination of the entire Mediterranean at stake.

The Opponents

- From the foundation of Rome in 753 B.C. until the Punic Wars about 500 years later, the Roman military was not noticeably better than its opponents in terms of training, professionalism, or equipment. In fact, it had suffered a number of defeats at the hands of various enemies:

  - After breaking away from Etruscan domination, Rome was sacked by Gauls in 390 B.C.

  - In 321 B.C., the Romans were beaten and ritually humiliated by the Samnites at the Battle of the Caudine Forks.

  - In the 280s, the Greek mercenary general Pyrrhus destroyed two Roman armies.

  - Finally, in the Punic Wars, the Romans lost several successive fleets and were out-generaled and defeated no fewer than three times in Italy itself by the Carthaginian military genius Hannibal. Rome's darkest hour came at the Battle of Cannae when, in one afternoon, Hannibal obliterated two entire Roman armies, killing between 60,000 and 80,000 men.

- Yet Rome usually ended up winning wars. The key to the Romans' early success was a dogged determination never to give up, no matter what the cost, coupled with vast reserves of manpower drawn from the Italian cities they had conquered and given citizenship. These manpower reserves repeatedly enabled the Romans to keep fighting and wear down their opponents.

- The Romans learned from their enemies, as well. They adopted weapons and tactics that took the best from each foe: for example, the short sword of the hill tribes of Spain, which evolved into the *gladius* and became an immediately recognizable symbol of Roman imperialism and military might.

- In its weapons, tactics, and organization, the army of Philip V of Macedonia was a direct descendant of the army commanded by Alexander the Great. At Alexander's death, his empire had split into near-constant fighting among rival kingdoms, with each using Alexander's style of warfare. These Hellenistic kingdoms were powerful and wealthy, and collectively, they controlled the richer, more urban, more culturally sophisticated eastern half of the Mediterranean.

Roman and Macedonian Forces and Technology

- By the time of Cynoscephalae, a new Roman army, with better weapons, better tactics, and more training, was in place. In its earliest phase, it probably fought in something like the phalanx used by the Greeks, but by the late 4th century B.C., it had begun to use a system sometimes called the manipular army.

© Getty Images/Photos.com/Thinkstock.

Around the time of Cynoscephalae, the Roman military made the transition from a primarily citizen militia to a professional standing army.

- In this system, the army was drawn up in subunits called maniples, blocks of 120 men arranged into three lines on the battlefield, in a chessboard-like formation.
    - Those on the first line, the *hastati*, wore a helmet and had a large oblong or rectangular shield called a *scutum.* They were armed with javelins and the *gladius*. The second line, the *principes*, were similarly equipped. The third line, the *triarii*, had longer thrusting spears and may have been composed of older men.
    - There were also groups of lightly armed skirmishers called *velites* and a contingent of cavalry.
    - In combat, the maniples used a loose formation that allowed soldiers in the back ranks to come forward and replace those in the front row, thus keeping fresh those actually engaged in the fighting.

- A major advantage was the way the army was subdivided into units of steadily decreasing size. Roman soldiers and officers were trained to fight and maneuver in any of these size increments, and the soldiers were trained to quickly shift the direction of the fighting. This made for maximum flexibility in battle, in direct contrast with the massive Hellenistic phalanx, which while powerful when lumbering forward for a direct assault, was slow and could not easily be subdivided or redirected.

- The heart of Philip's Macedonian army was its phalanx, composed of a block of 15,000 men armed, just like Alexander's phalanx, with the long sarissa. They were supported by about 10,000 cavalry, missile troops, and skirmishers.

The Generals

- Philip V was an able and experienced general and politician, sometimes likened to Alexander. By the time of the Battle of Cynoscephalae, he had held the throne for more than 20 years and was well regarded as a leader. He had led a number of successful military campaigns, extending the borders of his empire into the islands of the Aegean, and had skillfully attempted to create a coalition of states to oppose the growing power of Rome.

- Titus Quinctius Flamininus also had considerable military experience before Cynoscephalae and had distinguished himself as a junior officer in the Roman army during the Second Punic War.
    - One distinguishing feature of Flamininus was that, for a Roman, he was unusually infatuated with Greek culture. He spoke fluent Greek and was an avid collector of Greek art.

    - This proved useful when fighting Philip because Flamininus was able to persuade a number of Greek states to join him by presenting himself as the savior of Greece, who would free it from Macedonian control.

The Battle

- Cynoscephalae represented much more than a clash between Rome and the Hellenistic kingdoms of the east. It was a battle between two different types of military systems. The two sides were almost exactly evenly matched at about 25,000 men each; thus, the battle would be a good test of which system was superior.

- The battle opened with some confused skirmishing between small detachments, indicating to both generals that their opponent's main force was nearby. Flamininus accordingly deployed his army in the customary formation: the three rows of maniples screened by the *velites*. By choosing to move immediately into combat formation, Flamininus gained an advantage: He would begin the battle with his troops arranged exactly as he wanted them. On the other hand, deploying immediately meant that his men had to line up on the slope of one of the hills, thus yielding the advantage of the higher ground to Philip's army.

- By the time Flamininus's men made contact with the enemy, the Macedonian phalanx had formed up and could add the momentum of a downhill charge to its already formidable strength. The Roman left could not resist the weight of this attack and began to give ground, retreating back down the slope. Even though they were being steadily pushed back, it was a fighting retreat, and the Romans kept their formation and did not panic.

- Flamininus had a few war elephants with him with which he opened gaps in the Macedonian lines—exactly the sort of weak points that the Roman manipular system was designed to exploit. Reeling from this assault, the Macedonian left began to run, pursued by the Romans.

- This was the crucial point of the battle. A substantial gap now developed between the two halves of the respective formations, effectively splitting them into separate battles. The Romans were in the process of winning one side, and the Macedonians the other, with overall victory still up for grabs.

- It was at this key moment that the flexibility of the Roman system showed its value. A junior Roman officer ordered a number of maniples from the right wing to break away, wheel 90 degrees to the left, and attack downhill against the rear right of the Macedonian phalanx that was threatening to overwhelm the Roman left. The effect was immediate. The men of the cumbersome Macedonian phalanx were unable to move to meet this new threat. Attacked from both the front and behind, the Macedonian right disintegrated, and the men were slaughtered as they attempted to run away.

Outcomes

- Cynoscephalae ended the war. Following the victory, Flamininus announced that the Greeks were now free from Macedonian oppression. What the Greeks failed to realize was that the Romans now regarded them as their clients, and they had to do whatever Rome wanted.

- For the moment, the Romans allowed Philip to retain his throne, but within two decades, another war broke out.
  - If there were any lingering doubts about the superiority of the Roman military system or any idea that the outcome of Cynoscephalae had been a fluke, the Battle of Pydna put them to rest in a Roman rematch against the Macedonian phalanx. Pydna was a head-on clash in the sort of environment in which the phalanx usually excelled, yet it crumbled.

  - What Cynoscephalae had shown and Pydna confirmed was that the well-disciplined, determined, and flexible Roman military machine was now qualitatively superior to most of its foes.

Suggested Reading

Connolly, *Greece and Rome at War*.

Hammond, "The Campaign and Battle of Cynoscephalae in 197 BC."

Plutarch, *Life of Flamininus*.

Polybius, *History of Rome*.

Questions to Consider

1. In what ways was the Roman legionary system superior to the Macedonian phalanx?

2. Can you identify any potential weaknesses in the Romans' style of fighting?

31 B.C. Actium—Birth of the Roman Empire

Lecture 6

As early as the 3rd millennium B.C., naval warfare occurred in the Mediterranean, and by the 2nd millennium B.C., the Egyptians were conducting large-scale amphibious warfare, transporting troops in ships from the Nile region and landing them in Palestine. The Battle of Actium is representative of the dominant style of naval combat used for most of history; it marks the end of the Roman Republic and the beginning of the Roman Empire, as well as the solidifying of Rome's domination over the entire Mediterranean basin.

Ancient Naval Warfare

- During a nearly 4,000-year stretch beginning in the 3rd millennium B.C., the main warship was the galley, a long, narrow vessel propelled by dozens or even hundreds of rowers. Such warships lacked the stability or the storage capacity to travel the open seas and typically hugged the coasts.

- One battle technique was to equip these vessels with a ram at the bow, which they would try to crash into the hull of enemy ships. The Greeks of the 5th century B.C. emphasized seamanship, with the primary goal to position a ship so as to ram the opponent broadside. A variant on this tactic was to sweep alongside an enemy vessel and break off its oars.

- All these maneuvers demanded speed, nimbleness, and a high degree of skill from the rowers. The ship design that developed to optimize these qualities was the trireme. A classic trireme was about 115 feet long and only 20 feet wide, with some 200 rowers on three levels packed into the narrow hull. A triple-pointed bronze ram weighing 400 pounds was affixed to the bow of the ship, and a small number of marines, usually no more than 35, stood on the upper deck armed with bows.

The *Olympias*, a modern replica of a trireme, proved surprisingly fast in trials, achieving speeds of more than nine knots for short bursts.

- The Romans preferred fighting on land. Accordingly, rather than using the ramming strategy, they tended to favor naval battle tactics that involved ships meeting in such a way that the soldiers they carried could fight each other.

- Naval warfare had already been tending this way with the construction of larger warships; they could carry greater numbers of soldiers and were less maneuverable, but they often had collapsible wooden turrets from which several men could shoot bows or even throw down rocks on an enemy's deck. They also began to carry catapults and ballistas that could hurl stones or oversize arrows at the crews of opposing ships.

Historical Background and Opponents

- In the 150 years following Cynoscephalae, the Roman Republic swallowed up the remaining Hellenistic kingdoms and extended its boundaries around almost the entire coast of the Mediterranean

Sea. Although Rome had achieved great success with its overseas conquests, these same successes had created severe internal strains in the fabric of the republican system of government.

- What had been a ruling coalition of powerful families had devolved into a few strong men dominating through wealth, power, and prestige. During a violent five-decade stretch beginning around 100 B.C., a sequence of these men—Marius, Sulla, Pompey, and Julius Caesar—had fought a brutal series of what were essentially civil wars. Yet none was able to hold on to power permanently.

- Even if the republic was no longer a political reality, many of its ideals remained powerful, and the strongest of these was a deep-seated aversion to being ruled by a king or anyone who acted like a king. This sentiment led to the assassination of Julius Caesar in 44 B.C., when his behavior became too monarchial.

- Caesar's death left two candidates vying for power in Rome:
    - The first was Caesar's second in command, Marcus Antonius, who inherited most of Caesar's wealth, prestige, and the loyalty of the majority of his legions. Mark Antony was direct and tough, a military man who got along well with the common soldier; he could hold the allegiance of his men by sharing their witticisms and living conditions, but he did not have the subtlety of mind for complex political machinations.

    - The second was Octavian, a boy just out of his teens, whom Caesar posthumously adopted as his son. Octavian possessed a brilliant, coldly calculating mind and a flair for manipulating affairs to give the appearance he desired them to have. He also had a gift for finding men who were talented where he was deficient, such as his childhood friend Agrippa, blunt and forthright, a natural military commander of considerable genius. With Octavian directing grand political strategy and Agrippa compensating for his deficiencies on the battlefield, they were a formidable pair

- Because neither Octavian nor Antony was quite ready for open conflict, they divided the Roman world between them while each continued to maneuver for dominance. Antony chose the eastern half of the empire, which had always been the richer portion. Octavian was left with west, but this included Italy and Rome itself.

- Antony's sphere also encompassed the last remaining independent state, Egypt, now ruled by the young Queen Cleopatra, a highly intelligent and assertive woman. Their relationship provided the wily Octavian with the ammunition for a war of propaganda against Antony, whom he portrayed as totally under Cleopatra's control. Octavian's rumor mill cleverly exploited traditional Roman phobias of domination by foreign monarchs.

The Battle

- In 32 B.C., Antony moved his and Cleopatra's main army and fleets to the western coast of Greece. Most of the fleet was put into the Gulf of Ambracia, where it would be safe from storms, and the bulk of the army was encamped nearby. Expecting Octavian to attack from the north, Antony also occupied a number of key towns along the coastline and fortified others to the south, guarding his supply lines to Egypt.

- Agrippa, in charge of Octavian's campaign, demonstrated his cleverness with his first move. Rather than attempting to land in Epirus in the north, as Antony expected, he led a naval assault against the city of Methone far to the south, threatening Antony's vital supply line, and began raiding other points on the southern coast of Greece.

- Falling into Agrippa's trap, Antony diverted many of his ships from the northern coast to the south. Octavian now brought the main body of his army across from Italy and landed them at Panormus in the north, just as Antony had expected, and they marched down the coast to the Gulf of Ambracia. A stalemate developed, with Antony's army encamped on the southern side of the gulf and Octavian's on the northern.

- Meanwhile, Agrippa continued his raids along the coast, capturing one strategic position after another. This string of victories not only raised the morale of Octavian's troops while eroding that of Antony's, but they also put a stranglehold on Antony's supply lines. Compounding his problems, Antony's camp was on low-lying ground, and his men were wracked by malaria and dysentery; as a result, Antony's army began to suffer deaths and desertions.

- Antony now had to act before he lost his whole force, and he chose to fight Octavian at sea with about 230 ships, 20,000 legionaries, and 2,000 archers. It is uncertain whether Antony's plan was to attempt to defeat Octavian's fleet or merely to break through and escape with as much of his force as possible.

- Antony deployed his ships in four groups—a right wing, a left wing, a center, and a reserve—each consisting of about 60 galleys, with himself in command of the right wing. Octavian's forces of nearly 400 ships and about 40,000 soldiers took up position in a broad arc, with Agrippa on the left side, facing Antony, and Octavian on the right.

- Antony had hoped that Octavian would rush forward to attack him in the shallows, where Octavian's greater number of ships could not all fit. Agrippa refused to be drawn in and waited for Antony to come out to him, even backing up slightly to ensure that Antony would have to fight in open water, where Agrippa could envelop his flanks. The two forces came together all along the line, and the battle devolved into a series of individual combats among small groups of ships. Two or three of Octavian's smaller ships clustered around one of Antony's, first exchanging arrow and ballista fire, then closing to board.

- As the battle wore on and as Octavian's ships sought to outflank Antony's, a gap opened up in the center of Octavian's line. Immediately, Cleopatra's reserve squadron raised their sails and lunged through. Making no attempt to engage Octavian, they set course for Egypt. Antony, seeing Cleopatra escaping, transferred from his flagship to a small, fast craft and set off after her.

- Antony's forces continued to fight bravely for a considerable time but were overwhelmed. Estimates are that around 150 of Antony's ships were taken or destroyed, added to the approximately 150 already defeated during the campaign leading up to the final battle. Antony's fleet was annihilated, along with most of the best troops of his army.

Outcomes

- Antony may have believed he could recover and rebuild his forces after the Battle of Actium, but his reputation was irretrievably damaged. Although he caught up with Cleopatra and both reached Egypt safely, in terms of opposing Octavian after Actium, they were finished; within a year, both committed suicide.

- Octavian was now the sole ruler of the Roman world. Drawing on his skill at manipulating images, he cloaked his power in innocuous-seeming titles and accrued to himself the powers of the main offices of the Roman state while cleverly refusing to hold the offices themselves, lulling Romans into a false belief that the republic lived on. In reality, it was dead, and Octavian, under the new name of Augustus, became the first Roman emperor.

- Augustus's settlement of the Roman state and his establishment of the position of emperor would form the system of government for the next 500 years. He and the empire would continue to exert a powerful influence as role models for a long line of future leaders, from Charlemagne to Napoleon to Mussolini.

Suggested Reading

Casson, *Ships and Seamanship in the Ancient World.*

Gurval, *Actium and Augustus*.

Morrison, et al., *Athenian Trireme*.

Sheppard, *Actium 31 BC*.

Questions to Consider

1. Why do you think naval combat centering on oared galleys remained so constant in the Mediterranean for so many thousands of years?

2. What key mistakes did Anthony make that contributed to his defeat at Actium, and which one was the worst?

260–110 B.C. China—Struggles for Unification

Lecture 7

Few countries can boast that they have had approximately the same culture, language, borders, and religions continuously for more than 2,000 years, but such is the case with China. During a roughly 150-year period, from 260 to 110 B.C., a series of strong leaders and generals, through military force, welded together a group of separate and highly antagonistic kingdoms to create the unified state of China. They founded the long-lived and influential Han Empire, the model for all subsequent Chinese dynasties, and finally, they fought off attacks by a menacing tribe of central Asian nomads to define China's northern frontier and establish control over the territory that would develop into the economically vital Silk Road.

Chaotic Times in Chinese History

- The entire later history of China was forged in the wars of the period from 260 to 110 B.C., yet it was not a single battle that marked these turning points; in each case, they were determined by longer military campaigns or even a series of wars.

- The heartland of China is a rough square measuring about 2,000 miles a side. This land is divided into northern and southern halves, each centered on one of the great basins of the Yangtze and Yellow rivers and each with its own distinctive climate and terrain. By 1700 B.C., large, culturally sophisticated empires, such as the Shang dynasty, had emerged along the Yellow River.

- During a tumultuous 500-year period from 722 to 221 B.C., the core region of China fragmented into rival kingdoms that fought one another fiercely for dominance. The first half of this turbulent era is called the Spring and Autumn period, while the second half is the Warring States period. The famous military text *The Art of War*, attributed to Sun Tzu, was written in and seems to reflect the style of warfare used in these chaotic times.

Chinese Armies and Technology

- Chariots played a central role in early Chinese warfare. They were technologically advanced, with excellent wheels that were dish-shaped rather than flat and made with up to 32 spokes. These qualities made them stronger, especially when taking sharp turns. By using parallel shafts and breast straps around the horse's chest, rather than a pole-and-yoke arrangement and a neck strap, as in many Western chariots, Chinese versions were also more efficiently attached to the horses.

- The cab was large and typically held crew of three: a charioteer in the middle, an archer to one side, and a soldier armed with a halberd on the other.
  - This halberd was a distinctively Chinese weapon with a long shaft topped by a large bronze head that combined several cutting edges and spikes, one of which was at right angles to the shaft.

  - This weapon was held sticking out one side of the chariot parallel to the ground, where it would slice into infantry or potentially even sweep an enemy charioteer from his vehicle. Used in this way, it has no equivalent in Western warfare, wherein the main weapon of chariots was the bow.

- The era of chariot warfare in China witnessed some massive battles. For example, at the Battle of An in 589 B.C., at which the army of the state of Jin defeated their counterparts from Qi, the Jin forces included some 800 chariots.

- The reign of the chariot on the battlefield in China began to draw to a close with the introduction the crossbow around the beginning of the 4th century B.C. Powered by a cranking mechanism and able to fling a bolt more than 200 yards, the crossbow was a great leveler on the battlefield because it did not require years of training to master, had a greater range than a bow, and could bring down elite charioteers or their horses at long distances. Soon, all Chinese armies incorporated large contingents of foot soldiers armed with crossbows.

- At around the same time, horses began to be bred larger and could be ridden; thus, the first cavalry units began to appear in Chinese armies. Much more mobile and cost-effective than chariots, cavalry gradually took over the roles formerly played by chariots.

Qin Unification

- In the 3rd century B.C., the Qin, one of the many squabbling kingdoms, finally emerged to dominate and defeat all the rest, resulting in the unification of all China, north and south, for the first time.

- No single decisive moment marked the ascension of the Qin. Between 364 and 234 B.C., the armies of Qin fought at least 15 separate campaigns against their rivals in what amounted to an ongoing war of attrition. In 221 B.C., this series of conquests was

Discovered in a series of burial pits near Xian, the terra-cotta warriors—approximately 8,000 life-size statues—preserve an invaluable portrait of the Qin army.

completed by the man who became the first emperor of China, Qin Shi Huangdi.

- The Chinese army of this era was primarily an infantry force made up of troops with several types of weapons.
  - The backbone of the infantry was issued the distinctive Chinese dagger-axe, supplemented by a bronze sword as a secondary weapon.
  - Flanking these heavy infantry formations were other contingents of foot soldiers carrying crossbows, who would fire mass volleys of bolts into the enemy.
  - For armor, the better-equipped troops wore conical metal helmets and body armor made by lacing together small plates of toughened leather or metal. Most arms and armor were fashioned of bronze.

- Cavalry were mostly archers equipped with powerful compound bows.

- Officers were career professionals who could rise through 17 successive ranks, beginning with the officer in charge of a unit of 5 men. Promotion depended on demonstrating skills, performance in battle, and mastery of texts on military theory, such as Sun Tzu's famous *Art of War*.

- The Chinese system offered rewards for those who personally killed a certain number of enemies in battle, and officers in charge of groups larger than 100 could earn promotion or rewards for the total number of enemies slain by the men under their command.

The Han Empire

- The Qin Empire collapsed almost immediately upon the death of Shi Huangdi, but he had set the vital precedent of a united country that could not be erased. The group that stepped up to establish a stable, long-lasting empire in China would be the Han dynasty, and they, too, would come to power by force of arms.

- By 209 B.C., two main rivals had emerged. The first of these was Hsiang Yu, who represented the powerful southern kingdom of Chu. His opponent, representing the Han kingdom, was Liu Pang. With a constantly changing cast of allies, these two men fought a series of wars over the next several years, culminating in the Battle of Kai Hsia in 203 B.C.

- Hsiang Yu began the battle by leading a charge straight into the larger army of Liu Pang but found himself surrounded and his forces in danger of encirclement. After a day and a night of fierce fighting, he was forced to retreat and establish a fortified camp that was immediately besieged by the Han-led army.

- Hsiang Yu determined that his best chance was to break out with as many of his elite troops as could be saved. He spent the night drinking and saying farewell to his wife, then mounted his favorite horse and led 800 of his cavalry in a desperate charge.

- The Han seem to have been taken by surprise; Hsiang broke through their lines and headed south, pursued by 5,000 of the Chu cavalry. After receiving some misdirections from a farmer, however, he rode into a swamp, and the Chu riders caught up and began picking off his men.

- Eventually, he was trapped on a hilltop with only 28 men left. Heroically leading his men in a counterattack, he managed once again to break free, killing more than 100 Chu enemies plus a general in the process.

- Hsiang now reached the banks of the Yangtze River, where a boatman offered to ferry him across to safety, but perhaps realizing that his fortunes would never recover, he refused, and he and his remaining companions turned to face the pursuing Han. Fighting until he was wounded several times, in a final act of defiance, he cut his own throat.

- With the death of Hsiang, Liu Pang completed his conquest of the main kingdoms. He now took the title of emperor and changed his name to Kao-ti, becoming the first member of the Han dynasty.

Challenges to the Han

- The Han dynasty was challenged by a dangerous confederation of northern nomads. These were the Xiong-nu, a group of tough and militant horse archers from the area of what is today Mongolia.

- In 200 B.C., Kao-ti led an expedition against the Xiong-nu that turned into a disaster. His army was surrounded by a horde of Xiong-nu. Unable to deal with the threat by military means, Kao-ti had to adopt a policy of appeasement, paying tribute to the nomads and providing hostages.

- In the aftermath, the Han undertook military reforms that placed a greater emphasis on the use of cavalry so that they could counter the mobility of the Xiong-nu with their own rapidly moving forces.

- By 130 B.C., the Han launched a series of aggressive raids against the Xiong-nu. The emperor who oversaw these campaigns was Wu-ti; his subordinates managed to push back the Xiong-nu in a series of campaigns and secure the northern frontier. Because of these campaigns, China would permanently control the territories through which ran the eastern section of what would become known as the Silk Road.

Li Ling against the Xiong-nu

- In 99 B.C., the Chinese general Li Ling led 5,000 infantry into Mongolia, but through an apparent miscommunication, his supporting cavalry force did not arrive. Li Ling continued to advance and was surrounded at the Tien Shan Mountains by 30,000 Xiong-nu horsemen.

- Undaunted, the Chinese infantry closed ranks, with men holding spears and shields in the front and crossbowmen behind. Wave after wave of charging Xiong-nu attacked this formation, but the Chinese held together, with the crossbowmen slowly but steadily shooting down the enemy cavalry. Finally, the Xiong-nu broke off, leaving thousands of their dead on the battlefield.

- Li Ling than organized his men in a mobile defensive formation and began the long march back to the Chinese border. Another army of at least 30,000 Xiong-nu now joined in and, for nearly a week, swarmed around the beleaguered Chinese, but with a stubborn series of rear-guard actions, Li Ling held them off and approached the frontier with about half his force remaining.

- Now the Chinese had to traverse a narrow gorge, and the Xiong-nu were able to block the route and roll boulders down on the Chinese, finally breaking their formation. Only 400 of the Chinese made it safely across the border. At the last minute, Li Ling either was captured or voluntarily defected to the enemy, among whom he lived for the rest of his life.

Suggested Reading

Cotterrell, *Chariot.*

Peers, *Soldiers of the Dragon.*

Portal, ed., *The First Emperor.*

Sawyer, *Ancient Chinese Warfare.*

Questions to Consider

1. Who do you think would have won, and why, if the ancient Chinese Qin army had fought the Egyptians? What about the Greeks, Macedonians, and Romans?

2. What qualities of the Xiong-nu made them such a persistent threat to China?

636 Yarmouk & al-Qadisiyyah—Islam Triumphs

Lecture 8

A 50-year-old Arabic woman named Hind stood at the edge of a camp on an arid field in Syria. As she peered into the swirling dust, she began to feel that something was terribly wrong. Throughout the day, individual soldiers, most of them wounded, had come seeking aid for their injuries from the battle taking place a few hundred yards away. But now, healthy warriors had panicked and were fleeing the enemy. Hind rallied the other women who, brandishing tent poles, flinging stones, beating drums, and singing songs, accused the fleeing warriors of cowardice. Shamed and humiliated, the men turned and reengaged the Byzantines, pushing them back and reestablishing their lines.

Competing Empires

- Pressure from nomadic Germanic tribes, combined with a number of other factors, resulted in the political collapse of the western Roman Empire and its fragmentation into a host of barbarian kingdoms, many of which still continued to emulate the culture and model left by Rome.

- Meanwhile, the eastern half of the Roman Empire—the Byzantine Empire—had continued to flourish under an unbroken string of emperors and controlled territory from Constantinople across all of modern Turkey and along the eastern coastline of the Mediterranean, even including Egypt.

- Further east, the Sassanid Empire encompassed eastern Syria and Mesopotamia and extended through the Caucasus into south central Asia, reaching all the way to the borders of India. Culturally and geographically, the Sassanians, a version of the old Persian Empire, created a golden age for Persian culture, and the empire was wealthy, vast, and powerful.

The Opponents

- The Byzantines and Sassanids fought a series of wars over the vital crossroads territories of Armenia and Syria. At the beginning of the 7th century, the Sassanian king won substantial territories, only to lose most of them to a counterattack by the Byzantine emperor, Heraclius.

- During the next decade, a new power, within the span of a single year, inflicted a pair of stunning and decisive defeats, first on the Byzantines, then on the Sassanids.

- This new force was Islam, and within a century, Muslim armies would achieve one of the most impressive conquests of all time. More important, it would be one whose effects were among the longest lasting and influential on the modern world.

The Battle of Yarmouk

- The Battle of Yarmouk was the culmination of several years' clashes between Rashidun and Byzantine forces along the Mediterranean coastline. The Muslims had won most of these battles, and Heraclius realized that they were a serious threat.

- Deciding to wipe out the invaders, Heraclius assembled an army of 30,000 to 80,000 men under the overall command of an experienced Armenian general named Vahan. This army encamped in a strong position on a rocky plateau surrounded by steep but shallow gullies near the Yarmouk River.

- The Rashidun had been seeking a decisive battle; thus, their various small armies gathered into a force estimated at 15,000 to 30,000 men—no more than half the size of the Byzantine army.

- Tactical command of the Rashidun went to Khalid Ibn al Walid, an excellent tactician, who often used light cavalry to make dramatic flanking moves that struck his enemies from unexpected directions.

- While the two armies confronted each other across the Yarmouk plain for nearly three months, reinforcements kept arriving on the Muslim side, eventually persuading the Byzantines to attack before they lost their heavy numerical advantage.

- Vahan drew up his forces in four large blocks stretching out over perhaps as much as five miles.
    - The main body of each section was composed of a mass of infantry with a cavalry unit positioned behind. Vahan posted his best heavy infantry on the right to serve as anchor for the line. Vahan took up his place in the center.

    - In addition, Vahan had a large number of Arabic auxiliaries and cavalry led by their own officer.

- Khalid also organized his men into four large blocks, each opposing one of Vahan's. Like the Byzantines', the blocks were mainly infantry with small cavalry units behind them. However, Khalid also kept a large central reserve of cavalry to the rear under the command of an especially bold officer named Zarrar.

- On the first day of battle, as was customary, a number of single combats occurred, in which champions from each side stepped out from the ranks and challenged one of the enemy to a one-on-one fight. Seeing that the battle of the champions was going against him, Vahan ordered some of his infantry forward, and skirmishing occurred along the line.

- The second day opened at first light with a much more aggressive push by the Byzantines, again along the entire front. Vahan's strategy seems to have been to engage the main forces of the Rashidun in the center and drive forward on the right and left wings to envelop and surround them.

- As the day wore on, the plan started to work, but each of the four main sections of the Rashidun army had located its own camp directly behind the line, and as the right and left wings retreated, they entered this zone of camps, where they encountered their infuriated wives.

- To shore up the crumbling left and right, Khalid now brought forward his cavalry reserve, first on the right, then the left. Between the rallying of the Rashidun infantry and counterattacks by the cavalry, the Byzantine advance was pushed back. By nightfall, the original battle lines were restored.

- On the third day, the Byzantines concentrated their offensive on the northern end of the battlefield. Again, they met with initial success, pushing the opposing Rashidun formations backward into their own camps, and again, the women refused defeat. Once more, Khalid's timely insertion of the cavalry reserves resulted in the Byzantine offensive being blunted and forced back to its original line.

- Vahan began the fourth day with another strong, generalized assault, which again had initial success. But when Khalid counterattacked with his cavalry reserve, the Byzantine infantry and cavalry lost contact. The cavalry were driven off to the north, exposing the infantry to harassing attacks by Khalid's horsemen. The Rashidun were able to take control of the northern end of the field and push the Byzantines back.

- When a gap opened in the Byzantine lines, the dashing Zarrar and a small cavalry contingent drove into the rear lines of the Byzantine army, where they seized control of a bridge that constituted the only good crossing point over the treacherous gullies and ravines around the river. Zarrar not only trapped the Byzantine army but also cut off its supply line.

- On the fifth day, Vahan attempted to negotiate for a truce. Khalid, however, sensing that the initiative was swinging in his favor, refused. These negotiations took up most of the day, and Khalid used the time to reorganize his remaining cavalry into one large strike force.

- The final day of the battle opened with Khalid launching his offensive. His infantry forces surged forward along the length of the field; meanwhile, his massed cavalry corps swept around the northern end of the field and down against the left flank of the Byzantine lines. Vahan attempted to counter with his own cavalry but was too slow in deploying them.

- As the Byzantines began to retreat, they backed into a funnel-shaped peninsula, and the constricted units began to panic and lose cohesion. The defeat turned into a rout. The brilliant Khalid had played the Byzantines perfectly, blunting their brute-force assaults with his well-timed cavalry charges until they became exhausted, then striking back with a single, powerful blow that won the battle.

Outcomes of Yarmouk

- Yarmouk was a truly decisive battle. After it, the Byzantines made no further major attempts to oppose the Rashidun armies in Syria and basically ceded the entire eastern Mediterranean to them. All of Egypt, Palestine, and Syria easily fell to Muslim forces within a few years.

- The Byzantines retreated into Anatolia and, later, to the walls of Constantinople itself. Behind those massive fortifications, they held out for another 1,000 years, but their broader empire was gone.

- Yarmouk was the moment when the future of the Middle East was determined. Until that point, the dominant culture had been Greco–Roman. Today, all the countries in the region (with the exception of Israel) are predominantly Arabic-speaking and Muslim.

The Battle of al-Qadisiyyah

- Shortly after Yarmouk, the Battle of al-Qadisiyyah in southern Iraq, an equally decisive conflict, was fought with similarly long-lasting effects.

- The Rashidun army was about 15,000 to 30,000 men, among them 5,000 veterans of Yarmouk.

- The Sassanian army of 30,000 to 60,000 boasted a number of especially dangerous units, including a substantial corps of trained war elephants imported from India. The Sassanians were renowned for their excellent cavalry, especially a group of elite heavy cavalry who were covered from head to foot in scaled metal armor and whose horses were protected with metal scales, as well.

- Whereas Yarmouk lasted six days, al-Qadisiyyah was a five-day battle, but the general pattern of the Muslim line fending off a series of enemy frontal charges day after day was the same. Finally, the Sassanian commander was killed, and on the fifth day, the Rashidun forces broke the Sassanian formation and won the battle.

- Just as Yarmouk resulted in dramatic and permanent changes to the culture, language, and religion of the eastern Mediterranean, so, too, did al-Qadisiyyah. Centuries of domination by the Zoroastrian religion were wiped away, replaced with Islam. The old Persian language and culture were blended with new Arabic elements.

- It is not an overstatement to say that, in cultural, linguistic, and religious terms, the map of the modern Middle East was drawn in the early 7th century A.D. The present (and the future) of the region was forged on the battlefields of Yarmouk and al-Qadisiyyah.

Suggested Reading

Donner, *The Early Islamic Conquests*.

Farrokh, *Shadows in the Desert*.

Haldon, *Byzantium at War, AD 600–1453*.

The History of al-Tabari, vols. 11 (Blankinship, trans.) and 12 (Friedmann, trans.).

Nicolle, *Yarmuk 636 AD*.

Questions to Consider

1. What key factors contributed to the success of the Arabic armies at Yarmouk and al-Qadisiyyah?

2. Do you think Yarmouk or al-Qadisiyyah was the more decisive battle, and why?

751 Talas & 1192 Tarain—Islam into Asia

Lecture 9

Sports fans and military history enthusiasts share an obsession for arguing about the outcome of hypothetical confrontations between opponents from different times or places. For military history buffs, these arguments often focus on such questions as: What would have happened if an ancient Chinese army had fought Alexander's Macedonians? One major battle was actually fought between a Chinese army and a western one—the Battle of the Talas River in 751 A.D., which matched an army of the Tang dynasty against an Arabic army of the Abbasid caliphate. The Battle of Tarain was another interesting clash of widely differing civilizations and empires—in this case, a 12th-century invasion of Hindu Rajput India by a Turkish Islamic army.

The Tang Dynasty

- The Tang dynasty constituted one of China's golden ages. From the early 7th century and for the next 300 years, the Tang ruled from their capital city of Chang'an (modern Xi'an), which became a great cultural metropolis, larger and more technologically sophisticated than its European counterparts.
    - By the early 700s, Chang'an had a population approaching a million and may have been the largest city in the world at that point.
    - It was an unusually cosmopolitan city, reflecting the far reach of the Tang. As the terminus of the Silk Road, it was a node of trade in whose vast markets goods from east and west were bought and sold.
- The Tang pursued an aggressive expansionist policy, and China's borders stretched west into central Asia and south into Vietnam. Along the Silk Road, they had established control over the Tarim Basin.

- On a modern map, the area of the key conflicts is roughly where Kazakhstan, Kyrgyzstan, and Uzbekistan overlap to the east of Tashkent. In the late 740s, this was the zone of operations of an enterprising Chinese general named Gao Hsien-Chih, the Tang commander at the Battle of the Talas River.

The Abbasids

- Meanwhile, a new power was making its presence known in the area: the Abbasid caliphate. Much like the Tang dynasty, the Abbasid caliphate was an aggressive, expansionist empire experiencing a cultural flourishing.

- Under the Abbasids, the city of Baghdad became a great center of culture, much like Chang'an. In 750, the establishment of Baghdad as the Abbasid capital was yet in the future, but the Abbasid armies were pushing ever eastward toward central Asia and India.

- The Arabic armies still possessed strong momentum from their initial wave of conquests of the eastern and southern Mediterranean. In some sense, conflict between the Tang and the Abbasids seems inevitable: they were both new, vigorous empires, looking to expand their influence and power.

Battle of the Talas River

- The incident that led directly to the Battle of the Talas River concerned some petty central Asian kingdoms. The rulers of Ferghana and Tashkent were feuding, and the king of Ferghana sought assistance from the local Chinese official, Gao Hsien-chih. Tashkent, in turn, sought aid from the Abbasids, and both the Chinese and the Abbasids responded by dispatching armies.

- The Abbasid general, Zihad ibn Salih, deployed his troops in standard formation, with a line of archers in front, the spear men drawn up behind them, and the heavy cavalry to the rear. In addition, there were lightly armed skirmishers and a cavalry reserve.

- According to Chinese sources, the turning point in the battle occurred when a group of Karluk Turks, who had been fighting on the side of the Chinese, switched sides during the battle and attacked the Chinese from the rear. Gao attempted to hold his fleeing troops together, but most were slaughtered or captured, and supposedly only 2,000 returned to China.

Significance

- The Battle of the Talas River set the high-water mark of Tang expansion to the west, effectively marking the end of Chinese attempts to project influence beyond the Tarim basin.

- At the time of the battle, the people of central Asia followed a variety of religions, but over the next few centuries, they became almost exclusively Muslim. Some historians have argued that the battle was decisive not only because it stopped the spread of Chinese power westward, but because it determined the permanent religious orientation of the entire area.

Rajput versus Turk

- Conversion to Islam after the Battle of the Talas River was a process that took several hundred years. Among the groups that converted were the warlike Turkish tribes of the steppe, who were outstanding horsemen and archers, and in the 11th century, several energetic Turkish leaders used these martial skills to carve out substantial empires.

- The most famous of these was Mahmud of Ghazni, who built an empire based in what today is Afghanistan and extending into Iran and Pakistan. Mahmud led 16 raids into northern India, where his zeal for destroying Hindu temples and statues earned him the nickname "Idol Breaker."

- Mahmud's empire was supplanted by a pair of brothers from Ghor. One established himself on the throne of Ghazni, while the other, Mohamad of Ghor, looked southward to India to carve out his own kingdom.

- In northern and central India, the Rajput kingdoms dominated and frequently fought each other. Rajput warfare emphasized large masses of conscript infantry coupled with war elephants. They did use cavalry, but this was a lesser arm of the military and was hampered by lack of the excellent horses available across the mountains.

- One of the prominent northern Indian Rajput kingdoms was ruled by a man named Prithvi-raja, an efficient and ruthless leader who controlled a large territory based around the cities of Ajmer and Delhi. The main invasion route into India from Afghanistan led directly through Prithvi-raja's territory; thus, in the late 12th century, Prithvi-raja and Mohamad of Ghor were on a collision course.

The First Battle of Tarain

- In 1191, Mohamad led a large army south into India and seized the fortress of Bhatinda. Prithvi-raja summoned his conscripts and moved to intercept Mohamad. The armies met at a site called Tarain, just north of Delhi.

- The available sources claim that Prithvi-raja's army numbered 200,000, with 3,000 elephants, and that Mohamad's was considerably smaller. The actual numbers are probably less than half of these, but it does seem that Mohamad's army was much smaller, although more professional.

- Prithvi-raja's forces charged enthusiastically forward, seeking to crush and envelop Mohamad's army by sheer weight of numbers. Mohamad was wounded severely enough that he had to leave the field, prompting his troops also to withdraw. The First Battle of Tarain ended as a victory for Prithvi-raja, but not a decisive one.

- It is interesting to compare how the two leaders responded to this battle.
    - Mohamad organized a new army with a greater emphasis on mobile horsemen and discipline. He purged his army of any

who had shown cowardice, increased training, and obtained more horses.

- In contrast, the only action taken by Prithvi-raja to prepare for the rematch was to secure agreements from other Rajput rulers to assist him, ensuring that his army would be even larger.

The Second Battle of Tarain

- In 1192, Mohamad returned, and Prithvi-raja moved to meet him with an army said to be more than 300,000 strong. They met on almost the same spot, but this battle would have a definitive outcome.

In the aftermath of the First Battle of Tarain, Prithvi-raja showed complacency, assuming that his forces had already proved their superiority.

- Mohamad divided his army into five units, including four with 10,000 light cavalry each, armed mainly with bows. The last was a division of 12,000 heavily armored horsemen. His plan was clever:
    - The four horse archer divisions were to advance and retreat repeatedly, all the while raining arrows down on the Rajput ranks. If attacked, they were to fall back and not engage in close combat.
    - If one division was being pursued with determination by a unit of the Rajputs, it should retreat, but the other three archer divisions should harass the flanks of the Indians until they gave up the pursuit.

  - With their inferior cavalry and slower horses, the Rajputs would not be able to chase their opponents very far, and Mohamad's horse archers would once again turn, charge, and fire more arrows.

- The plan took advantage of the strengths of Mohamad's army and exploited the weaknesses of the Rajput. It was also a classic confrontation between a smaller, highly disciplined force that used hit-and-run tactics and a large, unwieldy one that relied on brute force and numbers.

- Finally, Mohamad ordered a general retreat, his men pretending that they were panicking and fleeing. The undisciplined Rajput troops took the bait and rushed headlong after their apparently running foes, losing cohesion in their own ranks in their eagerness to get at their tormentors.

- This was the moment Mohamad had been waiting for; he ordered into battle his reserve division of 12,000 heavily armored horsemen. They smashed into the disordered Rajput ranks and began slashing with their swords. The other four divisions wheeled back to join the slaughter, and the battle became a bloody rout.

Outcomes

- With this victory and several follow-up campaigns, Mohamad of Ghor took control of many of the cities and kingdoms of northern India, including Ajmer and Delhi, establishing a permanent Muslim influence and community in these regions and a dynasty of Muslim monarchs in Delhi that would rule until they were deposed by Mongol invaders in 1290.

- The Second Battle of Tarain was a decisive one in world history because it secured a permanent Muslim presence in India. For the next 800 years, up to the 21st century, there would be tension and sometimes outright war between Hindus and Muslims.

- When India achieved independence from British rule in 1948, these tensions resulted in the establishment of Muslim Pakistan as separate from predominantly Hindu India, a national rivalry that continues to have significant effects on global politics even today.

Suggested Reading

Dunlop, "A New Source of Information on the Battle of Talas or Atlakh."

Joglekar, *Decisive Battles That India Lost.*

Peers, *Soldiers of the Dragon.*

Sandhu, *A Military History of Medieval India.*

Verma and Verma, *Decisive Battles of India through the Ages.*

Questions to Consider

1. What do you think would have been the long-term historical consequences if the Tang had won the Battle of the Talas River?

2. How do you think the army of Mohamad of Ghor would compare with the other armies we have studied thus far in the course?

1066 Hastings—William Conquers England

Lecture 10

In England during the 11th century, it was common for famous people to be given nicknames, often visually descriptive, to distinguish them from others with the same name. Thus, one man was called Harold Fairhair; another, Harold Bluetooth. Sometimes, the nicknames indicated aspects of their character or personality, such as Robert the Pius. Perhaps the most unfortunately named was William the Bastard, a son of Robert the Magnificent, duke of Normandy. William, perhaps sensitive about his nickname, crossed the English Channel in 1066 with an invasion force, won the Battle of Hastings, and conquered England. His victory not only changed the course of English and world history, but it also earned him a more flattering nickname: William the Conqueror.

English Succession Problem

- In England, 1066 began with the death of King Edward the Confessor. His demise created a succession crisis, out of which four claimants emerged:
    - Harold Godwinson, a relative of Edward's wife, was from a powerful family that controlled much English territory, and Edward may have made a deathbed acknowledgment of him.
    - The king of Denmark traced his descent from Cnut, an earlier king of England.
    - King Harald Hadrada of Norway was eagerly looking to extend his power over more territory.
    - William of Normandy was supposedly promised the English throne by Edward the Confessor when William visited in 1052.
- Several sources assert that when Harold visited Normandy in 1063, he swore an oath that he would recognize and support William as heir to the English throne.

- In the short term, Harold took the throne of England, but Harald Hadrada and William of Normandy were determined to contest his claim by force of arms.

The Opponents

- Harold summoned his supporters and retainers, the professional core of his army. He also called out the fyrd, a general summons somewhat like a militia. By late summer, Harold had gathered these forces in southern England near the coast, in anticipation of William's landing.

- Meanwhile, William began mustering his troops and constructing ships, but weather and other factors delayed him. By early September, Harold's men were running low on provisions and the mandated tour of duty for the fyrdmen was expiring. He had to begin disbanding his army.

- Just then, Harald Hadrada landed a force of 300 to 500 ships in northern England, moved inland, and captured York.

- Harold Godwinson acted swiftly, recalling his troops and covering 190 miles in just five days. On September 25, he came upon the main Norse army at Stamford Bridge.

- The Battle of Stamford Bridge was hard-fought. Hadrada was eventually struck down by an arrow that pierced his windpipe, but the Norsemen fought on even after their king was slain, bolstered by the arrival of reinforcements from their ships.

- In the end, the English prevailed, but both sides suffered heavy casualties. Some have claimed the Battle of Stamford Bridge as one of the key battles in world history, an argument that has some validity. This battle definitively ended the growing Viking influence in England that might have oriented England toward Scandinavia rather than the European mainland.

- A few days later, word came that William had taken advantage of Harold's absence to cross the channel and land his army, organizing his forces on a road that led to London in the peninsula near Hastings, where his army could be resupplied by sea.

- Harold marched south, reaching London in five days, where he paused for several days to assemble more troops. On October 13, Harold camped within a few miles of William, and both armies prepared to do battle the next day.

The Armies and Their Technology

- The arms, tactics, and equipment of Harold's Anglo-Saxon army and William's Norman army shared much in common:
  - The wealthier warriors on both sides would have worn long shirts of chain mail or scale armor with long vertical slits extending up from the bottom edge to allow horseback riding, often augmented by a mail coif that went over the head.
  - Both sides carried wooden shields with a metal boss and rim that were either circular or round on top, then tapered to a point at the bottom, a design especially handy for cavalry use.
  - Both sides favored straight, long, double-edged swords, designed for slashing attacks. Javelins and thrusting spears were also popular weapons.

The Bayeux Tapestry is one of our best sources of information regarding the arms and armor used by both sides at the Battle of Hastings.

- In the Bayeux Tapestry, men of both sides are shown wearing conical helmets, usually augmented with a nasal bar extending down from the helmet rim, although this style of helmet is more often associated with the Normans.

- Although the size of the Norman army is uncertain, the best guess is that there were about 2,000 horsemen, 4,000 infantry, and 1,500 archers and crossbowmen.

- A weapon popular among Anglo-Saxon warriors was the axe. It came in two varieties: a small, hatchet-like design that was thrown and a larger, heavy axe with an asymmetrical head whose bottom edge was longer than the top, used with two hands.

- Harold's force is estimated to have been slightly larger than William's and to have been composed of roughly one-third semi-professional warriors and two-thirds fyrdmen.

The Battle

- On October 14, 1066, William moved out with his men, heading inland, where he knew he would find Harold's army. Soon, the enemy had been sighted, and William deployed his men for battle.

- The topography of the battlefield favored the English: the top of a hill about 700 meters long whose sides were protected by gullies, ensuring that the English line could not be outflanked.

- Harold placed his most heavily armored men in front, forming a solid shield wall backed up by the lesser warriors and fyrdmen. This formation, thought to have been perhaps 10 men deep, stretched along the entire length of the ridge. Harold himself took up a position in the center, just to the rear of the phalanx.

- The Normans deployed on the low ground within about 200 meters of the English shield wall. In front was a thin screen of archers and crossbowmen, then a block of infantry, and the cavalry in the rear. William, like Harold, placed himself in the center, just behind the main lines.

- William commanded his crossbowmen to open fire on the English, but the Normans were firing disadvantageously uphill, and the English front row would merely have raised their shields until the barrage was over.

- William next ordered a general advance, and his line attacked the English shield wall. There was bitter close combat all along the line, with the English hacking with swords and axes while the Normans struggled to find or break a gap in the line of shields.

- Having been ordered to support the infantry, the Norman horsemen could not gain much momentum, but their role was to exploit holes in the enemy formation made by their infantry. Despite this new influx of forces, the English shield wall held firm against all attempts to penetrate it.

- The Norman left began to retreat down the slope. Some of the English fyrdmen, believing the enemy was routed, surged forward in pursuit. This was dangerous, exposing the center to a potential flank attack; thus, the Normans in the center also began to back down the hill. Even worse, at that moment, a rumor flashed through the ranks that William had been killed.

- Here was the crisis point of the battle: William's army was now on the verge of complete panic and flight. In response, he acted quickly and decisively.
    - William either removed his helmet or pushed it back so that his face was clearly visible, and his men could see that he still lived.

    - William's half-brother rode forward to rally the fleeing troops and perhaps bring reinforcements to the crumbling left flank.

    - William himself gathered a group of his knights and charged with them into the mob of advancing Englishmen.

- The counterattack was effective, and the English troops were cut off and slaughtered.

- William ordered his troops back up the hill, and the close-range fighting resumed. His cavalry also returned, adopting a new strategy of charging and pretending to retreat, in the hope of drawing the English out of their organized formation on the ridge.

- The ranks of the English were beginning to thin, with many of the more experienced and better armed soldiers having been killed or wounded and their places in the front rank taken by the less well-equipped militia and fyrdmen. William began one more major effort with a barrage of arrows and another general charge up the hill.

- This time, the arrows had a more deadly effect on the now inferior English troops, and one arrow apparently struck Harold in the face or eye. Accounts differ as to whether this wound was immediately fatal, but he dropped to the ground.

- Meanwhile, William's men advanced around the edges of the depleted English line so that the English were fighting on three sides. Both of Harold's brothers were slain, and he was finished off by slashing blades. With their deaths, the line finally shattered, and the Normans crested the hill and slaughtered the remnants.

Outcomes

- On Christmas Day, William was crowned in Westminster Cathedral, officially becoming king of England. Ironically, Harold's earlier victory at Stamford Bridge solidified William's position by both eliminating the Viking threat and winnowing the ranks of those who might have opposed him.

- The Norman Conquest blended Anglo-Saxon and Norman culture and reoriented England from Scandinavia to the European mainland.

- In particular, it strengthened the ties between England and France and ensured that England would be embroiled in continental affairs.
- It could even be argued that the Battle of Hastings marked the beginning of England's rise in world affairs, culminating in the 19th-century British Empire.

Suggested Reading

Bradbury, *The Battle of Hastings*.

Bruce, *The Bayeux Tapestry*.

Gravett, *Hastings 1066*.

Questions to Consider

1. Who do you think had the most legitimate claim to the English throne: William the Bastard, Harold Godwinson, Harald Hadrada, or the king of Norway?
2. To what degree do you believe that William's victory was due to the lucky arrow hit that killed Harold?

1187 Hattin—Crusader Desert Disaster

Lecture 11

Thirty years after the Battle of Hastings, Pope Urban II gave one of the most influential speeches in history at the Council of Clermont. His words sparked a conflict that stretched over several hundred years and involved all the major powers of Europe and the Near East. This conflict, of course, was the Crusades, a series of invasions of the Islamic kingdoms of the Near East by the Christian kingdoms of Europe. Although the Crusades extended over a 200-year span, the turning point was the Battle of Hattin in 1187. The opposing forces seem evenly matched, but squabbling among their leadership and a series of poor decisions ultimately doomed the European army to a miserable and dusty death.

The Crusades up to Hattin

- The enthusiastic response to the Pope's call for an expedition to free the Holy Land from the "pagans" was probably much greater than he imagined. Around 30,000 professional fighting men from a number of countries volunteered.

- Enlisting was called "taking the cross" because strips of cloth in the shape of a cross were sewn onto enlistees' clothes or painted on their shields. It was from this practice that they became known as Crusaders and the entire movement was called the Crusades.

- The official First Crusade set off in 1096, reaching Antioch in 1097. After a siege, Antioch fell, and the Crusaders marched down the coast, winning a string of victories. Eventually, they reached Jerusalem itself, and after another siege, in July of 1099, the Crusaders burst into the holy city and massacred many of the inhabitants.

- A number of small Crusader kingdoms were established along the eastern coast of the Mediterranean, the most important of which was the Kingdom of Jerusalem.

- From the European perspective, the First Crusade had been a spectacular success, exceeding all of its goals in just a few years.

- One significant effect was the establishment of Christian military religious orders, the Knights Hospitaller and the Knights Templar, which grew into powerful, independent political and military forces whose members took religious vows but were trained as knights. They quickly acquired a large number of key castles and strongholds scattered around the various Crusader kingdoms and, because they represented a permanent, professional core of fighting men in the region, became important factors in the overall history of the Crusades.

Motivations for joining the Crusade have been widely debated; some Crusaders hoped to acquire territory in the Holy Lands, while others were drawn by the chance to earn remission for their sins.

- The politics of this era are a confusing muddle of ever-shifting treaties between and among the small Christian kingdoms, the various Islamic states and their subfactions, the militant orders, the European kingdoms, and the pope.

- Periodically, a new wave of Crusaders would arrive from Europe, whose allegiance to either the existing Crusader states or their home countries was questionable and who often did not feel constrained by existing treaties or alliances.

- The lines of military and political command were often muddy or disputed. Rivals might temporarily join forces to achieve a desired end, but almost all major campaigns were fought by uneasy coalitions with divided and often antagonistic leadership. This lack of a clear command structure was a serious flaw that repeatedly caused problems for the Crusaders and hampered the effectiveness of their military initiatives.

- This stalemate continued for nearly 100 years. Then, in the late 12th century, a new Islamic leader emerged who would begin to tilt the balance in favor of the Muslims. He is known in the West as Saladin or, more correctly in Arabic, Salah-huddeen. The turning point of the Crusades, and one of Saladin's greatest victories, was the Battle of Hattin in 1187.

Leaders and Armies

- Saladin played an active role in the struggle for control over Egypt and eventually became its de facto ruler. From this base, he extended his power north, bypassing the Crusader kingdoms along the coast but gaining control over Damascus and much of Syria and Palestine.

- Saladin's armies were a mixture of Muslim Turks, Arabs, and Kurds, most of whom were fairly professional soldiers, with a high proportion of skilled horsemen. Standard equipment included bows, lances, spears, and swords.

- Many of Saladin's warriors were well protected with metal helmets, mail hauberks, substantial shields, and lamellar armor. At Hattin, Saladin probably commanded a force roughly equal in size to that of the Crusaders—around 30,000.

- In 1186, the Crusaders faced a leadership crisis.
  - Baldwin IV had appointed Raymond of Tripoli, an able commander, to serve as regent and rule the kingdom after his death. A captive for eight years in Aleppo, Raymond was fluent in Arabic, knowledgeable about Islamic civilization, and willing to explore peaceful coexistence.

- Guy de Lusignan was the choice of a group of nobles who favored a hostile approach to the Muslims and staged a coup at court in which Raymond was ousted.

- Reynald of Chatillon, like Raymond, was a skilled military commander and had been a captive in Aleppo for many years. The experience had left Reynald an implacable foe of the Muslims and a forceful member of the pro-war faction.

- Making up the majority of the Christian forces were various men-at-arms, both mounted and on foot, equipped with an array of weapons and armor, ranging from leather jerkins to mail hauberks, and bearing swords, axes, spears, bows, and polearms.

The Hattin Campaign

- In 1187, a truce was in effect between Saladin and the Kingdom of Jerusalem, but Reynald continued to raid Muslim caravans of merchants and religious pilgrims—plainly a breach of the truce. Saladin began assembling an army.

- One contingent of Saladin's army annihilated a Christian force at the Springs of Cresson that included several hundred Hospitallers and Templars. Although this battle itself was not decisive, it had a significant effect on the Battle of Hattin because the heavy losses incurred deprived the Christians of some of their most dependable forces.

- The various Christian factions temporarily shelved their differences and united under King Guy of Jerusalem. It was an uneasy alliance, with much bickering and resentments among the leaders. Guy ignored Raymond's advice against marching to relieve Saladin's siege of the city of Tiberias; Raymond had pointed out that the road to Tiberias lacked both water and fodder for the horses.

- The army traveled in several separate divisions, but each division assumed a similar formation, with the cavalry in the center, surrounded and protected by a hollow square of infantry. King

Guy was with the center group, which carried the most valuable Christian religious artifact, a cross alleged to be the one on which Christ was crucified.

- The soldiers literally baked in their armor, choked on the dust clouds raised by thousands of marching feet, and suffered tremendous thirst from the severely restricted water rations. Adding to the discomfort were harassing raids by Saladin's forces, which increased in intensity over the course of the day.

- By the afternoon, Guy determined to alter the direction of the march toward the nearby springs of Hattin, beyond which lay the Lake of Tiberias. Knowing that keeping the Christian army from reaching either of these water sources would be a great advantage, Saladin ordered a detachment of his more mobile army to hurry around and block the road.

- In contrast to the dismal conditions in the parched Christian camp, Saladin's army settled down for the night in high spirits.
    - Fresh sheaves of arrows were distributed to all the archers, and 70 camels carrying more arrows were organized to supply fresh ammunition wherever it might be needed.

    - A relay of more camels brought water from Lake Tiberias in goatskin bags so that everyone had plenty to drink.

    - Finally, Saladin had his troops arrange highly flammable dry undergrowth and sticks into bundles positioned along the windward side of the Christians' anticipated line of march.

- On July 4, the weary Christian army dragged toward the springs of Hattin, a couple of miles from the vanguard. To increase their discomfort and confusion, Saladin's men ignited the gathered brush, enveloping Guy's men in choking clouds of smoke.

- Saladin then launched a general attack. The battle now coalesced around the embattled Christian survivors atop the Horns of Hattin.

In the end, they were overrun, and most of the knights surrendered. The prisoners included nearly every major Christian leader.

Outcomes

- Hattin was a crushing defeat for the Crusader kingdoms. The main body of their fighting force had been destroyed and their leadership, lost. Not only was it a disaster in material terms, but it was also a devastating and humiliating psychological blow.

- Riding the momentum of his decisive victory, Saladin swiftly moved to capture many of the largest Crusader cities. Acre quickly fell, freeing 4,000 Muslim slaves and prisoners. By the end of the campaign, they would be joined by 15,000 more, while close to 100,000 Christians would be captured.

- Saladin began his siege of Jerusalem on September 20, using a sophisticated array of catapults and siege towers to bombard the city and its defenders. Seeing little hope for survival, the leaders of Jerusalem negotiated a surrender, agreeing to leave the city and pay a sizable ransom. On October 2, Saladin took possession. He would continue to rule until his death in 1193.

- Most of the history of the Crusades from this point on is a story of failures and attempts to cling to the cities remaining in Christian hands. In the end, all the Crusader kingdoms were destroyed and all Christian outposts in the region were lost.

- Although the Crusades seem to have been a fiasco for Europe, they helped to initiate the exchange of information and technology between East and West, an interaction that particularly benefited Europe and may even have helped prompt the Renaissance.

Suggested Reading

Kedar, ed., *The Horns of Hattin.*

Nicolle, *Hattin 1187.*

Riley-Smith, *The Crusades.*

Questions to Consider

1. Do you think the Crusades were ultimately doomed to failure because of logistics and other factors, or might the Crusaders have established permanent kingdoms in the Holy Lands?

2. What factors contributed to the Crusaders' defeat at Hattin, and which one was the most important?

1260 Ain Jalut—Can the Mongols Be Stopped?
Lecture 12

What do these warriors have in common: a tribesman wielding a blowgun and poisoned darts on the island of Java; a German nobleman trained to fight in heavy armor on a great war horse; a member of the fanatical Ismaili Muslim sect known as the Assassins, living in a mountain fortress in Syria; a Japanese samurai raised to follow the bushido code; a tough Afghan tribesman serving the sultan of Delhi in northern India; a Mamluk warrior of Egypt; a soldier of the Song dynasty in China; a Russian lord in Novgorod; and a Burmese war-elephant driver? The answer: Within just a few decades, all fought against the same enemy—the Mongols.

Great Conquerors

- The Mongols were arguably the greatest conquerors of all time. In three generations, they burst out of their homeland and swept across Europe and Asia, conquering every empire and civilization they encountered. By the end, their dominion stretched nearly 10,000 miles and constituted the largest contiguous land empire in history.

- The Mongols are notable for the astonishing diversity of their enemies, the range of environments in which they fought, and the different styles of warfare they outmatched.
    - The stereotypical Mongol warrior is a swift-moving nomadic horse archer, but many of their greatest successes resulted from mastery of siege warfare. From Syria to Korea, they captured walled cities and fortresses that were said to be impregnable by creatively employing a wide range of high-tech siege engines and weapons.

    - They built vast fleets and launched some of the largest amphibious invasions seen before the 20th century.

  - They were highly adaptable, quickly applying any new technology that seemed useful. When they had trouble breaking through the massive walls of fortified cities in China, they imported counterweight trebuchets from the west, manned by Muslim artillerymen.

  - When they attacked castles in Syria, they brought giant siege crossbows developed in China. From China, they also learned the use of explosives, which they employed both as weapons and as signaling devices on battlefields ranging from Burma to Europe.

- There was one notable instance, however, when a Mongol invasion was permanently stopped by a clear-cut defeat in open battle. This decisive battlefield victory was won by the Mamluk Egyptians at the Battle of Ain Jalut in 1260.

The Opponents

- The Mongol rise to power begins with the legendary figure of Temujin, a middle son of a tribal chieftain, trained to be an outstanding horseman and archer, able to endure harsh conditions and frequent deprivation.

- By adulthood, Temujin had accomplished the impressive feat of uniting the main Mongol tribes into a single fearsome horde. In recognition of his supreme leadership, in 1206, he received the name by which he would become mostly widely known: Genghis Khan.

- This leader organized highly efficient and mobile armies and dispatched them to subdue a neighboring rival state in northwest China and to attack the great central Asian power of the day, the Muslim Khwarazm Empire (based in today's Iran and Afghanistan), greatly extending Mongol power.

- The Mongol conquest of China started with the northern Jin dynasty, whose capital, Zhongdu (modern Beijing), fell in 1215.

Another Mongol army invaded Eastern Europe and wiped out an entire army of heavily armored Russian knights.

- Genghis Khan died in 1227, having appointed as successor one of his sons, Ogedei, who continued his policies.
  - Under Ogedei, the Mongols completed the defeat of Jin China, taking Jin's southern capital after an epic siege and moving on to invade Korea.

  - They began driving deep into southern China, ruled by the Song dynasty.

  - In central Asia, they conquered Georgia and Armenia and continued south into Kashmir and northern India.

  - They returned to Europe, overrunning much of Russia and Poland and moving toward Germany and Hungary. This attack was cut short when Ogedei died in 1241.

- If the Mongols were unusually tough warriors, their opponents at Ain Jalut, the Egyptian Mamluks, were similarly skilled. The Mamluk warrior class was an interesting phenomenon of the Muslim world, in which young slave boys, originally mostly of Turkic ethnicity, were raised in what amounted to military academies and trained to be highly professional and dedicated warriors.

- Despite their technical condition as slaves, Mamluks enjoyed fairly high status, and their officers often wielded considerable political power. During a time of internal turmoil in Egypt around 1250, they seized control, establishing the Mamluk sultanate. The third Mamluk sultan, the man who would confront the Mongols at Ain Jalut, was Qutuz.

The Prelude

- The origins of Ain Jalut can be found in the scheme of Mongke Khan (a grandson of Genghis) for world conquest. This plan involved dispatching Mongke's brother Hulegu to complete the subjugation of Persia, to eliminate the Assassins, and to obtain the submission of the main Islamic caliphates or, if they would not yield, to conquer them.

- Hulegu set out with proficiency and enthusiasm, accepting the surrender of various minor principalities, some of which supplied troops to augment his army. He besieged and captured dozens of the mountain fortresses of the Assassins, thought to be invulnerable because of their location.

- Hulegu then turned to the subjugation of the Islamic caliphates. The oldest and most prestigious of these was the Abbasid, based in Baghdad, whose current caliph believed that his religious authority would be enough to deter the invaders. In 1258, after a brief siege, the Mongols used their catapults to destroy one of the city's towers and poured into the breach.

- In 1259, Aleppo was taken by storm. Damascus capitulated soon after, and for all practical purposes, the Ayyubid caliphate toppled. Now the sole remaining major independent Muslim power in the region was the Mamluk sultanate of Egypt.

The Battle

- Just as Hulegu was preparing to take his army south and crush the Mamluk sultanate as he had crushed the Abbasids and Ayyubids, Mongke Khan died. Hulegu determined to return and participate in the selection of the next Great Khan, as was Mongol custom. He also decided to take the majority of his army with him. Mindful of his mission, however, he left behind a force of about 20,000 men under the control of his most trusted general, Kitbuqa, to mop up the remaining resistance in the area.

- Meanwhile, in Egypt, Qutuz, the Mamluk warlord, had been preparing his forces to confront the Mongols as soon as it became apparent that they were intent on invading. Now, with the withdrawal of the majority of the Mongol army, Qutuz saw an opportunity to attack the smaller contingent before reinforcements could arrive. He resolved to meet the Mongols before they reached Egypt, while their numbers were at a minimum.

- A wild card was the Crusader kingdoms. Although Jerusalem had been lost a half century earlier, the remnants of the kingdoms collectively constituted an appreciable force. They were the enemies of the Mamluks; accordingly, some Crusaders aligned themselves with the Mongols. But the majority were so alarmed by them that they officially adopted a neutral stance while covertly informing their old foes that they would be allowed to march through Crusader territory without opposition and even agreeing to help supply the Mamluk army.

- Qutuz augmented his forces through uneasy alliance with one of his former rivals, Baybars, who had gained a good military reputation fighting the Crusaders. Baybars was a gifted commander and an ambitious man; thus, for Qutuz, this alliance brought considerable risks. Yet together, the two men commanded a force roughly equivalent to, or even slightly greater than, Kitbuqa's Mongols. The two armies met at the springs of Ain Jalut in modern Israel, only about 15 miles southwest of where the Battle of Hattin had taken place.

- Baybars went ahead with an advance force and skirmished with the Mongols, sending word back to Qutuz to bring up the army. The battle took place on September 3, 1260. It appears that Baybars employed some of the deceptive tactics that the Mongols themselves typically used in battle. Over the course of the morning, the Mongols pressed aggressively forward, perhaps lured by feigned retreats.

- The Mongol assaults seem to have been especially heavy on the Mamluk left, and this section of Qutuz's army began to fall back for real and lose cohesion. They were temporarily rallied by a counterattack, but then the Mongols drove forward again and seemed on the verge of breaking through.

- Sensing a crisis point, Qutuz personally led another counterattack that succeeded in firming up the Mamluk left, and the two sides now closed in a deadly embrace. This must have been an especially fierce encounter, pitting against one another professional warriors who were similarly armed, used the same tactics, and were equally well-trained and experienced fighters.

- Sometime during this clash, Kitbuqa was slain or captured, executed, and the Mongols' line broke. The Mongol army split into several groups that met various fates. Some made a stand on a hill and were killed by Baybars; others fled into a swamp or into fields, where they were burned out; and a sizable contingent escaped to the north.

- Qutuz was not to enjoy his success for long. He entered Damascus in triumph and headed south to return to Egypt. Somewhere along the road, he was assassinated by a group of his own commanders. Baybars was appointed the new sultan. His heirs would complete the process of expelling the Crusaders.

The Mongols were likely somewhat surprised by the stubborn resistance of the Mamluk warriors, who were just as tough and well-trained as their opponents.

Outcomes

- Ain Jalut is a decisive battle for three reasons. First, it stopped the westward movement of the Mongols—a significant achievement, given that most of the Western powers had proven vulnerable to the battle tactics of the Mongols; it is not difficult to imagine Mongol armies rolling across Europe and the Mediterranean.

- Second, it ensured the survival of the Islamic states, which would rebound from the Mongol incursions to control most of the region.

- Finally, it shattered the image of the Mongols as unstoppable and terrifying warriors. Ain Jalut proved that even the mighty Mongols could be beaten.

Suggested Reading

Amitai-Preiss, *Mongols and Mamluks*.

May, *The Mongol Art of War*.

Morgan, *The Mongols*.

Turnbull, *Genghis Khan and the Mongol Conquests*.

Questions to Consider

1. Had the Mongol armies not repeatedly been recalled after the death of a khan, do you believe that they would ultimately have conquered Europe?

2. The Mongols based their armies on mounted warriors, in contrast to many of the armies of the ancient/medieval world, which focused on infantry. What are the advantages and disadvantages of each style?

1410 Tannenberg—Cataclysm of Knights

Lecture 13

On the morning of July 15, 1410, two armies drew up on opposite sides of a field in central Poland in what would be one of the largest battles ever fought in medieval Europe, involving over 50,000 combatants, many of them knights in full armor. On one side were the fearsome Teutonic Knights, supported by knights from Kulm, Konigsberg, and Brandenburg. The other side consisted of the combined armies of Poland and Lithuania. The subsequent Battle of Tannenberg would be a grim, violent affair that ended with the near destruction of one side and helped to determine the borders of Eastern Europe for many centuries to come.

The Teutonic Knights

- During the Crusades, a number of new monastic orders formed whose members took the usual religious vows but trained as knights. The most famous of these orders were the Templars and the Hospitallers, both of which were based in the Holy Land and played important roles in the history of the Crusades.

- A third major order of fighting monks, founded around the same time and originally based in Acre in the Holy Land, truly began to flourish when they established outposts in Europe. These were the Teutonic Knights; as the name implies, they were mainly Germanic, and they eventually focused their attention almost entirely on Northern and Eastern Europe.

- Their chosen opponents were the last remaining European pagans, found in Prussia, Lithuania, and other parts of northeastern Europe. The Knights set up a network of massive castles from which they launched their raids, and they aggressively pursued their chosen enemies, steadily expanding their territory and driving east. This expansion brought them into conflict with Poland, setting the stage for the Battle of Tannenberg.

- At the time of Tannenberg, the grandmaster of the order was Ulrich von Jungingen, from a noble southern German family. Over the years, he served in nearly every major position in the Teutonic Knights' hierarchy, including marshal of the order and commander of a castle. Although he was an effective military leader, contemporary sources describe him as headstrong, arrogant, and a bit impulsive.

The Teutonic Knights were professional soldiers encased from head to foot in the highest quality plate armor and mounted atop huge war horses specially bred to carry their enormous weight.

- The number of full-fledged ordained warriors equipped with complete suits of plate armor and mounted on great chargers probably amounted to only a few hundred. They were supplemented by a much larger number of Teutonic lay brothers, who trained as knights but were not ordained and fought as somewhat less heavily armed horsemen.

- Each castle also had larger numbers of hired men-at-arms, professional fighting men of lower social status. They mainly fulfilled the roles of light cavalry, foot soldiers, archers, and crossbowmen.

- Finally, the Teutonic Knights made extensive use of noblemen from many countries, who volunteered to serve temporarily as a demonstration of piety but were not part of the formal membership. Each of these guest Crusaders would have been accompanied by his own entourage of squires and fighting men.

The Poles and the Lithuanians

- In the late 14th century, the youngest daughter of the reigning king of Hungary was a girl named Jadwiga, who received an excellent education and spoke Hungarian, German, Polish, Latin, Bosnian, and Serbian. When she was 10 years old, through a complex series of events, she became heir to the throne of Poland and was crowned as king.

- Meanwhile, a succession crisis was going on in Lithuania, and great pressure was being exerted to convert this last pagan nation in Europe to Christianity. A young duke named Jagiello emerged as the dominant candidate. Jagiello had been raised a pagan, but he converted to Orthodox Christianity. Because he now wanted to marry Jadwiga, he agreed to become Roman Catholic.

- In 1386, the two married, and the countries of Poland and Lithuania were united; in 1387, Lithuania officially became a Christian country. Although many hailed the conversion of Jagiello as representing the final triumph of Christianity over the last European outpost of paganism, others, including the Teutonic Knights, dismissed it as insincere and, thus, to be ignored.

- Jagiello faced threats to his control of Lithuania, foremost from his cousin Vytautas, who had persuaded the Teutonic Knights to undertake several military actions against Lithuania, given that they refused to recognize Jagiello as a genuine Christian monarch.

- After several years, Jagiello and Vytautas agreed to shelve their differences and their rivalry and, instead, work together for the benefit of Lithuania. This somewhat unlikely-seeming alliance proved to be long-lasting and highly effective.

- In practical terms, Jagiello was the nominal ruler of Lithuania, but Vytautas oversaw day-to-day affairs. In 1399, Queen Jadwiga died, leaving Jagiello the king of Poland in both name and reality.

The Campaign

- The situation was volatile, and the spark that ignited it into open warfare was a rebellion against the Teutonic Knights by the inhabitants of a region called Samogitia in western Lithuania. With Ulrich taking a hard line in negotiations, Jagiello and Vytautas determined to launch an invasion of the Knights' territory.

- Lithuania hoped to recover Samogitia and Poland to acquire a lost province of its own, Pomerania. To achieve these aims, Jagiello and Vytautas decided to launch a strike northwards towards the Knights' main stronghold at Marienbad.

- As the allied army marched north, the Knights' army shadowed their movements, and eventually, the two forces drew together between the villages of Tannenberg and Grunwald. Most modern analysts believe that the army of the Teutonic Knights had 25,000 to 30,000 men, and the allied army of the Poles and Lithuanians, about 40,000 to 55,000.

The Battle

- On July 15, 1410, the two armies faced off across a shallow valley. The elite of the Teutonic Knights were deployed on the left side of the line, while the guest Crusaders constituted the right wing. On the other side, the Lithuanians formed the allied army's right wing, facing the Knights, while the Poles were deployed on the left.

- The combined Polish–Lithuanian army swept forward along the line, and the two sides crashed in a head-on conflict. This phase of intense hand-to-hand combat continued for nearly an hour, with neither side giving way.

- Then, abruptly, the Lithuanians and some of the allied forces on the right wing pulled back and apparently went into full retreat. This incident is controversial, with commentators both ancient and modern disagreeing widely as to its cause.
    - Some claim that it was a planned retreat designed to lure the Knights out of their formation and scatter them, in which

vulnerable condition they could then be counterattacked. This was a classic move used by steppe horsemen, and the Lithuanians were well acquainted with it from their experience fighting the Mongols.

  - Others maintain that it was a genuine retreat and that the Lithuanians were forced to fall back under pressure from the heavily armored Knights. Whether real or feigned, the allied army's right wing withdrew, and the guest Crusaders surged forward.

- Despite the collapse of the right, the Poles held firm in the center and on the left, battling the Teutonic Knights. Some of the Poles now even drove into the gap created by the advance of the guest Crusaders and turned to threaten the exposed flank of the Knights.

- At this crucial stage of the battle, Ulrich mounted his horse, gathered the reserve force of the Knights around him, and led a thundering charge in a wedge formation diagonally across the field and directly at the Polish royal eagle banner, under which he assumed he would find Jagiello.

- Had they broken through the Polish ranks and killed Jagiello, this charge might well have won the battle, but Vytautas saw it coming and, collecting a force of his best-equipped knights, moved to intercept. A fierce battle ensued, but the charge lost its momentum and faltered just short of Jagiello.

- The battle now turned decisively against the Germans. The remaining Knights were assaulted from the rear by Polish light cavalry, while on the right, the Lithuanians swept back onto the field, trapping the returning guest Crusaders between them and the victorious Poles.

- By the end of the day, between 10,000 and 15,000 on the German side were dead, with about the same number captured. It was a crushing defeat; the full-ranking Teutonic Knights were almost wiped out, with more than 200 lying dead on the battlefield, along with the entire leadership of the order.

Outcomes

- Although the Knights would survive for quite a while after Tannenberg, their power was much reduced and they went into a long decline. The Battle of Tannenberg helped to establish the borders of the states in Eastern Europe and effectively put an end to medieval German expansion into Poland and Lithuania.

- The battle is perhaps even more important in symbolic terms.
    - For the Lithuanians and Poles, it represents a high point of national pride and achievement, especially with regard to resisting invaders.

    - In World War I, when the Germans won a major victory over the Russians near the site, they named the later battle Tannenberg as well and represented it as having avenged the earlier loss.

    - Some later Germans romanticized the Teutonic Knights as a group nobly trying to bring Christianity and civilization to backward parts of Europe. Nazi Germany, for example, portrayed its own seizing of eastern territory as a continuation of the Knights' mission.

- The Battle of Tannenberg was the last major battle of the Middle Ages and was one of the last in which gunpowder did not play an important role. In terms of world history, it was the cusp of the vastly important Age of Exploration, when European seafarers begin to establish links with the rest of the globe.

Suggested Reading

Turnbull, *Tannenberg 1410*.

Urban, *The Teutonic Knights*.

———, *Tannenberg and After*.

Questions to Consider

1. How influential was the trio of Jagiello, Jadwiga, and Vytautas in the battle's outcome, and which one do you think made the greatest contribution?

2. Do you think that the battlefield utility of the knight in armor, as represented by the full-fledged Teutonic Knights, was worth the enormous investment in resources required to equip and maintain them?

Frigidus, Badr, Diu—Obscure Turning Points
Lecture 14

This lecture considers three unrelated battles involving very different participants and spanning more than 1,100 years. What links these battles is that each had important effects—in some cases, changing the course of global history—but information about the battles themselves is especially scanty or uncertain. This, then, is a lecture about decisive battles for which we can appreciate their long-term influence, even when relatively little is known about the course of the battle itself.

The Battle of the Frigidus River (394)

- In 312, Constantine defeated his rival Maxentius at the Battle of the Milvian Bridge to become emperor of the Roman world. Following a vision just before the battle, Constantine became the first Roman emperor to convert to Christianity.

- Throughout the 4th century, many Romans, especially senators, continued to worship pagan gods and to urge the return to polytheism, causing dissension between the eastern and western halves of the empire. The emperor of the eastern half was Theodosius, a fervent proponent of Christianity. The western empire was ruled by Valentinian II.

- When Valentinian was found dead under mysterious circumstances, the western empire passed into the control of a Frankish general named Arbogast, who appointed a new emperor, Eugenius.

- Eugenius was a well-known sympathizer with the Roman aristocrats who favored paganism, appointing a number of them to key government posts. Pagan shrines were restored, and it seemed that a pagan revival was underway.

- Theodosius was determined to quash this development; when diplomatic efforts failed, he organized his army for an invasion of the west, departing Constantinople in May 394 and crossing the Alps unopposed.

- The battle began with a headlong attack by Theodosius's men. Eugenius's troops stood firm, and the day ended in a standoff.

- On the second day, a tempest supposedly swept through the valley, with the high winds blowing directly into the faces of Eugenius's army. Our only sources for this battle are Christian ones, which emphasize the role of this storm, claiming that it was so powerful that it blew the arrows of Eugenius's men back at them. The army of Theodosius was victorious.

- The battle was perceived by contemporaries as a clear victory of the Christian God over the pagan ones, and it resulted in the deaths of many of paganism's most prominent adherents. Thus, Christianity was firmly established as the dominant religion, and the majority of the inhabitants of the Roman Empire converted.

The Battle of Badr (624)

- In 624, Islam had only a few hundred converts, who had been expelled from their hometown of Mecca and forced to flee to the nearby town of Medina, an event known as the Hejira.

- These converts were led by the Prophet Mohammed, struggling to spread his nascent religion and establish his authority. The expulsion from Mecca undermined this ambition and left him and his followers refugees. It seemed as if Islam might fade away before it started.

- Mecca was a major trading center and a key point on the caravan route. For this reason, Mohammed decided to raid the caravans, simultaneously providing a source of income and offering the satisfaction of getting back at the Meccans who had spurned them.

- After several years of minor raids, the tension between the early Muslims and Mecca came to a head when the Meccans organized an especially large trade caravan. This tempting target motivated Mohammad to make an attempt on the caravan. Anticipating this move, the merchants of Mecca likewise mobilized.

- The caravan managed to dodge Mohammad's force and reach Mecca safely, but the Meccans were determined to eradicate this menace to their trade. Mohammad likewise needed a successful battle in order to shore up his reputation.

- The two small armies eventually met at Badr, a spot in the desert where there were some wells. Mohammad reached Badr first and stopped up most of the wells, leaving only a few on some high ground; he then encamped his troops—about 300 men, 70 camels, and 2 horses—in a defensive position around them. The Meccans numbered more than 900, including several hundred cavalry on horses and camels.

- The battle began in a traditional manner, with champions from each side fighting duels. Then, the Meccans charged the Muslims on their knoll. The Muslims did not run to meet them, but instead responded with a rain of arrows on the Meccans as they labored up the slope.

- After some fighting, the Meccan attack faltered, and the Muslims surged forward, broke the Meccan line, and won the battle. About 70 Meccans were slain, with about the same number captured. On the Muslim side, losses numbered only 14.

- Badr is one of the few battles mentioned in the Koran, where the victory is ascribed to divine intervention. Whether one attributes the Muslims' success to this or to Mohammad's savvy generalship, Badr was a key turning point in the history of Islam.
    - It transformed Mohammed into a major political figure.

    - It added a military dimension to Islam, with Mohammad now acknowledged and respected as a victorious general and the

Muslims as an armed force to be reckoned with—one of the defining aspects of early Islam.

- It established Islam as a legitimate religion and infused its followers with a sense of confidence and pride.

The Battle of Diu (1509)

- One of the pivotal moments in world history was the 16th century, the era of exploration and colonization, when previously separate cultures and civilizations became aware of one another and were connected by new sea routes. For some civilizations, such as the indigenous peoples of the Americas, this contact would be disastrous, while for others, such as Spain and the Netherlands, it became the stepping stone to vast wealth and far-flung colonial empires.

- Most of the earliest European voyages of exploration were motivated by the desire to find a new seaborne path to India, Southeast Asia, and China, fabled lands of limitless riches—the source of rare spices, silk, and other prized luxury items.

- The reason for this search for a sea route was that all the land routes were controlled by Muslim powers that reaped most of the profits from the trade that passed through their territories.

The new deep-bellied ships of the 15th century were designed to provide the stability and cargo space required to traverse the oceans; such ships were soon also armed with rows of cannons along the sides.

- Islam had grown rapidly, and its dominion stretched from the Balkans deep into Africa and east to northern India. Fueling this expansion was the Islamic states' stranglehold on the lucrative Eastern trade routes.

- By contrast, Christianity was confined to Western Europe, hemmed in to the south and east by Islamic states. In technological terms, too, Europe had seemed to be stagnating during the Middle Ages, while such Muslim cities as Baghdad became centers of innovation and learning.

- Seafaring in the East at this time was similar to that practiced in the classical world, with the long, narrow galley propelled by hundreds of oars. Such vessels had an extremely limited range, rarely ventured out of sight of land, and could operate only in calm or enclosed waters.

- But here was one area of technology in which Western Europe had advanced: seafaring on the open seas. At the close of the 15th century, Europe had begun to produce deep-bellied, square-rigged ships capable of braving and even crossing the great oceans. This type of ship could also be armed by cutting holes along the sides and adding rows of cannons.

- Although many countries concentrated their efforts on sailing west, the seafarers of Portugal focused on an eastern route. They journeyed south around the tip of Africa, then explored the coastline back north toward the Arabian peninsula.

- By 1500, Portuguese mariners eventually found themselves in India. The Arabic merchants who controlled the trade were not pleased to encounter the Portuguese in their territory; seeing opportunity, however, the Portuguese seized several ports and raided Arabic shipping.

- The king of Portugal then dispatched a fleet of 21 ships under the command of Dom Francisco de Almeida, with orders that show a remarkably practical understanding of geopolitics and economics: “Nothing would serve us better than to have a fortress at the mouth of the Red Sea … because from there we could cut off the spices … and all those in India from now on could only trade through us.”

- Correctly perceiving Almeida’s expedition as a threat to their monopoly, an unlikely coalition of Mamluks of Egypt, Ottoman Turks, and ships from the local Indian rulers banded their naval forces together to oppose him. The Venetian Republic also felt threatened by the Portuguese and offered assistance.

- Almeida caught up with the combined fleet at the Indian port of Diu in 1509 and sailed boldly in to attack them. Almeida had 1,200 men on 19 oceangoing ships, 12 of which were of a carrack design and 7 of a caravel type.

- The combined fleet opposing him numbered more than 200 vessels: 80 to100 war galleys and the rest dhows and other small coastal craft.

- The battle was a wild melee, with the sturdy Portuguese ships firing broadsides from their rows of cannons while the swarms of galleys and other boats attempted to ram or run alongside and board. The battle was a disaster for the combined fleet, which was nearly entirely wiped out.

- Over the next several decades, there would be at least three more major naval battles between Portuguese ships and Muslim fleets along the northwestern coast of India, but all would end the same way.

- Diu marked the moment when Europe finally broke free of the crippling Islamic monopoly on trade with the East and began its steady rise to world domination. The economic shift resulting from the battle also set into motion the slow decline and eventual disintegration of the once-mighty Ottoman Empire.

- What the Battle of Diu had fundamentally determined was that the ultimate victor in the rivalry to control the rich trade with the East would be a Christian European power, not a Muslim one, an outcome that can truly be said to have profoundly shaped the rest of world history.

Suggested Reading

Cameron, *The Last Pagans of Rome*.

Diffie and Winius, *Foundations of the Portuguese Empire, 1415–1580*.

Krefft, *Ten Battles*.

Mathew, *History of the Portuguese Navigation in India*.

Weir, *50 Battles That Changed the World*.

Questions to Consider

1. Which of the three battles discussed in this lecture had the greatest influence on history, and why?

2. Does the success of Christianity owe more to the outcome of the Frigidus, or does the success of Islam owe more to the outcome of Badr, and would either religion have survived a loss at its respective battle?

1521 Tenochtitlán—Aztecs vs. Conquistadors
Lecture 15

The conquests of Mesoamerica and South America are among the most astonishing military stories of all time. For example, in one battle fought at the Inca capital of Cuzco, 190 Spaniards defeated an army of 40,000 Inca warriors with a loss of only one man. Similarly, in two years, Cortés, with fewer than 1,000 Spaniards, utterly destroyed the Aztec Empire. The subjugation of the Americas by Europe is one of the more controversial episodes in history. The conquistadors have alternately been lauded as brave men succeeding against all odds and condemned as rapacious invaders responsible for an appalling genocide. On the surface, these episodes seem to be dramatic demonstrations of European military superiority. How could these unlikely victories have transpired?

The Aztecs

- In 1500, the Aztecs were at the height of their power, yet only a few hundred years earlier, they had been a wandering tribe with no homeland, looked down on by nearly all other groups. In 1325, they settled on a swampy island in the middle of Lake Texcoco because of a divine prophecy that instructed them to make their home where they saw an eagle perched on a cactus eating a snake. The Nahuatl name for the nopal cactus was *tenocha*, and the city was dubbed Tenochtitlán.

- The Aztec society was both militant and theocratic, with priests and religion playing central roles. The Aztec pantheon of gods was a frightening collection, most of whom demanded regular human sacrifice. In Aztec belief, the world had been created by the gods using their own blood, and only regular offerings of human blood would enable it to continue.

- To meet the gods' insatiable demand for blood, Aztec warfare eventually became focused not so much on killing enemies in battle as on trying to immobilize and capture them so that they might later be ritually sacrificed.

Among the most feared of the Aztec warriors were the Jaguar Knights and the Eagle Knights, both of whom went into battle dressed in elaborate costumes resembling their animal namesakes.

- Not surprisingly, the Aztecs were not loved by their subjects, and they maintained their hold on power through fear and military might. The Aztec army was large and well organized. The elite soldiers were members of warrior fraternities who had repeatedly proven themselves in battle.

- Their main hand-to-hand weapon was a wooden club, lined on both sides with razor-sharp pieces of obsidian. The knights also carried small wooden shields, and their armor consisted of wooden helmets and quilted cotton body armor. Their helmets and armor were often coated in bright feathers or animal skins.

- These elite warriors were supplemented by ranks of less proven soldiers, similarly armed but without the elaborate adornment, and by large levies of less trained troops: archers with bows and arrows, slingers who threw stones with great accuracy, and men equipped with the *atlatl*, or dart thrower.

The Spaniards

- Between 1506 and 1518, some 200 Spanish ships traversed the Atlantic. They initially settled on the islands of the Caribbean and were headquartered in Cuba but, 20 years after Columbus, had still not ventured in force onto the mainland.

- In 1518, the governor of New Spain selected a minor nobleman named Hernán Cortés to lead an expedition to conquer Mexico. Cortés landed in Mexico in early 1519 with approximately 500 men. His soldiers were an unruly lot motivated by varying degrees of greed and piety. The majority were Castilian Spaniards, and most were already well-trained and experienced.

- All were equipped with high-quality Spanish steel swords and steel helmets that gave good protection to their heads. Many also had high-quality steel body armor, consisting of either solid breastplates or chain mail. Cortés also had some crossbowmen, whose weapons could accurately propel a deadly dart more than 200 meters, and a number of Spanish-Arabian war horses; the men who rode them were highly experienced and would play a key role in the battles to come.

- The army of Cortés also included some soldiers equipped with technologically advanced weapons: the arquebusiers, who carried an early type of gun; though heavy, awkward, and slow, this weapon could fire a powerful bullet. Cortés also had several light cannons. These were crude and small but still had a sizable shock value against those who had not previously encountered such weapons.

- Finally, Cortés had a pack of large, vicious, trained war dogs. They do not feature much in Spanish accounts of the expedition, but depictions of them leaping upon victims and rending their flesh are vividly drawn in the codices of the Aztecs, suggesting that they may have constituted a significant military asset for the conquistadors.

The Invasion of Mexico

- Cortés landed at Veracruz and began marching inland to Tenochtitlán. Because he had to leave some men to secure the ships and the coast, the force he led to conquer the Aztecs consisted of merely 300 soldiers. Of these, 40 had crossbows, 20 had arquebuses, and 15 were mounted on horses. In addition, he had three cannons and his pack of war dogs.

- Cortés fought and won several battles against the Tlaxcalan, the one major group in central Mexico who had not yet been conquered by the Aztecs. Eventually, they made peace and a formal military alliance, gaining Cortés tens of thousands of native troops.

- These initial battles showed that the steel armor of the conquistadors was virtually impenetrable by any of the native weapons. On the other hand, the Spanish swords easily sliced through the cotton and wood armor of the natives. Their crossbows and arquebuses were devastating, both from a distance and in crowds.

- The horses terrified the natives, who initially believed the horse and rider to be some kind of monster. Cavalry charges were devastating to formations of native troops, who had no counter to the assaults. Even the war dogs wreaked havoc.

- The Aztec emperor at the time, Moctezuma, seemed uncertain how to react and may even have believed that the appearance of the Spaniards marked the fulfillment of a prophecy. It was the height of the harvest season, when the Aztecs normally did not wage war; thus, Moctezuma invited Cortés to visit him at Tenochtitlán.

- On November 8, 1519, Cortés and his 300 companions entered Tenochtitlán; they were housed in a palace and treated as honored guests. After several days of sightseeing, Cortés kidnapped Moctezuma, taking him to the Spaniards' enclosure. The Aztecs did not know what to do; a tense stand-off ensued, during which Moctezuma was the "guest" of the Spanish.

- Now Cortés learned that 900 Spaniards had landed on the coast and that their commander, Narvarez, had orders to arrest Cortés and take over the expedition. Leaving only 80 men in Tenochtitlán under the command of Pedro de Alvarado, Cortés rushed back to the coast and rounded up some of the men he had left behind, amassing a force of 350.

- He entered into negotiations with Narvarez while secretly communicating with friends within Narvarez's forces and spreading bribes among Narvarez's troops. Cortés then launched a surprise night attack on Narvarez's headquarters. Narvarez and his lieutenants were captured, and through a mixture of bribery and skilled oratory, Cortés persuaded the rest of the soldiers to join him.

- Meanwhile in Tenochtitlán, Alvarado had been invited to attend a religious festival at which many high-ranking Aztecs were present. Perhaps seeing this as an opportunity to paralyze more of the Aztec leadership, Alvarado had broken the sacred peace and attacked the unarmed worshippers, slaughtering many of the Aztec aristocracy. Alvarado was now besieged in the palace by mobs of furious Aztec warriors.

- Cortés, to impress upon his troops the message that they had to succeed or die trying, ordered that the ships be destroyed. There would now literally be no turning back. Cortés managed to break through to Alvarado and join his forces, but the Spaniards were surrounded and besieged in the palace.

- The captive Moctezuma, who all along seems to have favored a conciliatory policy, agreed to urge the Aztecs to be calm. When he appeared, they stoned him, fatally wounding him. The new emperor, Cuauhtémoc, viewed the Spaniards solely as enemies to be exterminated and launched an all-out attack.

- On the night of July 1, 1520, Cortés and his army tried to flee across one of the causeways. Many of the conquistadors, in addition to their weapons, could not resist burdening themselves with the gold treasure, and as they tried to swim across the gaps in the causeways, hundreds drowned.

- Cortés escaped but lost half his army. As they marched away, the bedraggled survivors had to suffer the additional horror of watching their captured friends and comrades being dragged to the top of the main temple to have their hearts ripped out by priests and their bodies flung down the steps. This disaster became known as La Noche Triste, "the Sad Night."

- Cortés retreated to the territory of his Tlaxcalan allies and began planning the final assault on Tenochtitlán. It was during this gap that a new factor made its presence known: A smallpox epidemic broke out and swept through both the Aztecs and their allies.

- In 1521, Cortés returned to Lake Texcoco and began systematically capturing all the cities around its shore. He then constructed a fleet of 13 small ships, each equipped with a light cannon. These were used to seize control of the lake and cut off the causeways, thus preventing food and reinforcements from reaching the city.

- The Aztecs were driven back into Tenochtitlán, and Cortés and his allies laid siege. Under the leadership of Cuauhtémoc, the Aztecs refused to surrender, and Cortés had to invade. After months of bitter street fighting, by August 1521, the smoking ruins were finally in Spanish hands and the Aztecs had been virtually exterminated.

- The conquest of Mexico was a pivotal event: It opened up the Americas to European exploitation, with vast economic, cultural, and religious consequences, and it set the model for the era of European colonization that transformed the world.

Suggested Reading

Díaz, *The Conquest of New Spain*.

Hassig, *Aztec Warfare*.

Thomas, *Conquest*.

Robinson, *The Spanish Invasion of Mexico*.

Questions to Consider

1. Which of the following factors played the greatest role in the success of Cortés and why: disease, guns, steel, horses, attack dogs, native allies, Cortés's decisions, or Moctezuma's decisions?

2. Was the fall of the Aztec Empire inevitable? How might the Aztecs have successfully resisted the conquistadors?

1532 Cajamarca—Inca vs. Conquistadors

Lecture 16

In early 1527, two small Spanish ships crept tentatively down the unexplored western coast of South America. Just after crossing the equator, they encountered a large, well-made, oceangoing balsa raft. The crew of the raft was about 20 Inca merchants. This moment marked the first direct encounter between the Spanish and members of the vast Inca Empire. The Spaniards were impressed by the sophistication of the vessel's construction and even more excited by the silver and gold adornments worn by the crew. Such treasures were exactly what the Spaniards were seeking, and they immediately seized the raft and its contents. Within a few years, the mighty Inca Empire would fall to European invaders.

The Inca

- The Inca occupied a narrow strip of territory, several hundred miles wide but almost 3,000 miles long, stretching down the western coast of South America and encompassing parts of what are today Colombia, Ecuador, Peru, Bolivia, Chile, and Argentina.

- These regions had been home to a succession of indigenous civilizations and cultures that had been building urban sites at least as far as back as 1400 B.C. These civilizations developed a high level of craftsmanship in textiles, metalwork, and pottery.
  - They domesticated llamas and alpacas as sources of food, wool, and transportation.

  - They erected monumental structures using precisely cut stone blocks.

  - They developed religious beliefs and practices, such as sun worship, ancestor worship, mummification, and a sacred calendar.

- Some established large empires created by conquest and held together by an administrative structure and road-building.

- The Inca were relative latecomers to this environment, but all these elements would be incorporated into their culture. After founding the city of Cuzco, the Inca gradually increased in power until they controlled the entire valley and surrounding regions.

- Cuzco remained the capital city, but the empire was divided into four administrative districts of varying size. Cuzco was considered the center of the world, and four great highways leading to each of the regions converged at its central plaza.

- Rest stops and storehouses were erected at intervals along the roads so that the army could march swiftly, and a network of relay runners was established to carry messages. Altogether, about 25,000 miles of roads linked an empire with an estimated population of around 10 million.

- All Inca males were required to undergo basic military training, making it easy to raise armies of tens of thousands in times of crisis, and the system of storehouses greatly facilitated supplying these armies in the field. The soldiers were organized into groups of 10 and various multiples up to the largest unit size of 10,000, all foot soldiers.

- Popular missile weapons were slings and bolas—stones joined by a cord—that could be flung to entangle an enemy's limbs. The main hand-to-hand weapon was a club or mace consisting of a wooden shaft with a stone or metal head molded with pointy knobs. Armor consisted of quilted or padded garments, light shields, and helmets.

- In 1527, the first omen of disaster struck when a severe epidemic broke out, killing a significant percentage of the populace. Most likely this was smallpox. Among the victims was the emperor, as well as his chosen heir. Between the terror caused by the epidemic and the succession crisis caused by the deaths of the old emperor and his heir, the Inca Empire was thrown into confusion.

- In the civil war between the two remaining sons, Huascar and Atahualpa, Atahualpa prevailed. Apparently an able leader, Atahualpa had been able to call on the allegiance of the army and marched in triumph to the capital.

- It was precisely at this dramatic moment that Francisco Pizarro and his army of 167 Spaniards appeared, marching inland from the coast near where Atahualpa was encamped.

Francisco Pizarro managed to bring down the mighty Inca Empire with even fewer men than Cortés had commanded.

The Spaniards

- Pizarro was the illegitimate son of a Spanish military officer. Inspired by the tales of Cortés's success, he went to the New World and made several attempts to mount an expedition into unexplored southern areas. He entered into a partnership with several other men to outfit a pair of ships, and it was these that had encountered the raft of the Inca merchants in 1527.

- With the goods seized in this encounter as evidence, Pizarro returned to Spain, seeking royal backing for a major expedition. He got minimal financial support but was granted an official proclamation authorizing him to conquer Peru and naming him governor.

- Pizarro sailed from Panama on December 27, 1530. He proceeded slowly along the coast, eventually reaching Inca territory but encountering only ruins as a result of the devastation caused by the civil war between Atahualpa and Huascar.

- Finally, on November 8, 1532, Pizarro left the coast to head inland. He took with him 62 horsemen, 106 foot soldiers, and a few small cannons. This small army ascended up into the mountains, passing several points where their progress might easily have been halted by a defensive force.

The Campaign

- By extraordinary good fortune, Pizarro was making his advance just as the civil war was reaching its climax, and Atahualpa was encamped with an army of between 40,000 and 80,000 men at Cajamarca, near where Pizarro was marching into the highlands.

- Atahualpa sent an envoy to the Spanish bearing gifts and invited Pizarro to meet him at Cajamarca. He seems to have received accurate reports about their horses and weapons, but clearly, he viewed their numbers as too few to pose any threat.

- When the Spanish arrived at Cajamarca, they found it filled with the vast encampment of Atahualpa's army. The Spanish occupied low stone buildings lining the triangular main plaza. A Spanish embassy visited Atahualpa, with inconclusive results.

- Pizarro and his men began to fear for their lives, realizing that they were deep in Inca territory and isolated from any possible aid. They decided that their best chance of survival was to attempt to emulate Cortés's move and kidnap Atahualpa, using him as a hostage to ensure their safety.

- Accordingly, they invited Atahualpa to visit and prepared to seize him if the opportunity arose. The Inca leader promised to come. As the hours passed, the Spanish grew more agitated and sent a message promising that no harm or insult would befall him. As the sun began to set, Atahualpa, carried on a litter and accompanied by an estimated 7,000 of his chiefs and retainers, entered the village.

- Pizarro had his men, ready and fully armed, concealed in the buildings. He had also occupied a small fort and stationed his cannons and more men there. The Inca appear to have honored an agreement to come unarmed, although some may have carried slings and small knives.

- Eyewitnesses from the Spanish and Inca sides give somewhat differing accounts of what happened next, but what is certain is that Pizarro gave the signal to attack. His men, including 60 mounted on horses, burst out of the buildings and charged into the ranks of the unsuspecting Inca. The cannons fired with terrible effect into the crowded throng of natives, and the steel-encased Spanish began to cut them down. Pizarro himself led the attack on Atahualpa's litter.

- Oddly, the Spanish allowed Atahualpa to send and receive messengers and to act as the emperor while he remained captive. Atahualpa remained calm and dignified, clearly believing that he would soon escape. Noting the fascination that gold seemed to hold for the Spaniards, he offered to fill one of the nearby rooms with gold objects up to a height of about seven feet in exchange for his freedom. The Spanish eagerly accepted the terms, and Atahualpa gave orders for the gold to be collected and sent to Cajamarca.

- He badly misjudged the Spanish, however. Months passed while the ransom was collected, during which time the Spanish acquired reinforcements and took into custody the most powerful Inca generals, who might have organized opposition to them.

- Once the incredible ransom was assembled, the Spanish melted it all down, destroying an irreplaceable artistic heritage. Pizarro did not release Atahualpa as promised and, in July 1533, put him on trial on trumped-up charges and executed him.

- The Spanish replaced him with a succession of puppet emperors, while the rebel Inca elevated their own emperor. The last Inca emperor died in 1572, by which time Spanish consolidation of power over the Inca Empire was complete.

Outcomes

- The conquest and subsequent colonization of the Americas by European powers in the 16th century was certainly a turning point in history, with wide-reaching effects that are still felt around the world today. But the question remains: Why were the Spanish successful in the face of overwhelming odds?

- One obvious factor is naval technology. Developing the sort of oceangoing ships that could carry enough supplies, sail against contrary winds, and endure storms, along with the navigational instruments to steer an accurate course, was crucial.

- Another is the superior quality of Spanish steel, both in armor for protection and in swords for attack. This would definitely give a serious advantage, but there are many instances in history of groups armed with less advanced weaponry being able to defeat high-tech armies, even those that possessed gunpowder.

- Attention also often focuses on the Spanish horses, which were intimidating, and Spanish cavalry charges were initially devastating, but the Inca quickly learned to dig pits and take measures that helped limit the effectiveness of horsemen.

- Like most complex historical questions, the answer is still being debated, and the true explanation is probably some combination of factors. Nevertheless, the astonishing conquests of the Americas and collapse of the major indigenous empires constitute two significant military turning points in history.

Suggested Reading

D'Altroy, *The Incas*.

Hemming, *The Conquest of the Incas*.

McEwan, *The Incas: New Perspectives*.

Yupanqui, *An Inca Account of the Conquest of Peru*.

Questions to Consider

1. Which of the following factors played the greatest role in Pizarro's success and why: disease, guns, steel, horses, Pizarro's decisions, Atahualpa's decisions, or luck?

2. What fundamental similarities and differences are there between the campaigns and conquests of Cortés and Pizarro?

1526 & 1556 Panipat—Babur & Akbar in India

Lecture 17

Certain places seem fated to be battlegrounds—locations where major battles were fought, often hundreds of years apart. Usually, the reason is that the site constitutes a strategic crossroads, where invaders naturally would encounter defenders. Geographic determinism seems to be why the undistinguished little town of Panipat, about 80 miles north of Delhi, was the site of no fewer than three significant battles—in 1526, 1556, and 1761—each arguably decisive. The First and Second Battles of Panipat took place between local rulers and two of the earliest and most famous of the Mughal emperors, Babur and Akbar, and these battles established Mughal domination over the Indian subcontinent that would last until the arrival of the British.

Babur

- On June 8, 1494, in the fortress city of Akhsi, in the modern province of Ferghana, Uzbekistan, a local chieftain named Mirza was killed, leaving an 11-year-old son named Babur. Through his father, Babur could claim to be an heir of Tamerlane, and on his mother's side, he could trace his ancestry to the greatest of all Mongols, the mighty Genghis Khan himself. Despite his youth, Babur was already dreaming of becoming a great conqueror.

- Babur had a literary bent and, from an early age, wrote a diary that would survive and grow into a charmingly frank autobiography of his deeds, known as the *Baburnama*. In it, he describes how, after his father's death, he attempted to take control of Ferghana and succeeded in briefly capturing the important city of Samarkand.

- His youth and the treachery of others combined against him, however, and within three years, Babur had lost not only Samarkand but Ferghana and had been deserted by nearly all of his retainers. Undaunted, he rebounded from this low point, seizing the city of Kabul and making it the base for his operations.

- As his empire grew, Babur began to look to the riches of India and to lead raids into northern India. Since the battles of Tarain several centuries earlier, northern India had been dominated by the Muslim rulers of the Delhi sultanate.

- Babur determined to attack the sultanate, whose current sultan was Ibrahim Lodi, an Afghan Pashtun. The coming battle would be a study in contrasts, with one side emphasizing training and new technologies and the other relying on massive numbers and traditional methods.

The First Battle of Panipat (1526)

- Babur spent several years preparing for his invasion, assembling a highly professional army of about 10,000 soldiers. The core of his army was made up of excellent Turkic horsemen, well trained in the sort of hit-and-run wheeling attacks favored by steppe cavalry for centuries.

- Babur had acquired a cutting-edge military technology: gunpowder. The Ottomans had begun employing cannons and primitive guns to good effect, and by 1519, Babur had brought an artillery expert to Kabul to advise him in the use of this new technology. In addition to acquiring some cannons, Babur equipped a unit with matchlocks—primitive handguns that worked by thrusting a flaming wick against the firing hole to ignite the powder in the barrel.

- In February 1526, a scouting group from Babur's army defeated an advance element of Sultan Lodi's army. Babur now advanced with his main army to Panipat. Sultan Lodi had amassed a huge army to face the invaders. Sources claim a force of 100,000 men and 1,000 armored elephants.

- Babur ordered his 700 baggage carts tied together in a line, leaving wide enough gaps between them for cavalry to charge through. The matchlock men and light cannons were established on the carts and behind other temporary barriers, transforming the whole line into a series of miniature strong points. The remainder of the infantry was distributed around the carts, giving further protection to the slow-firing guns and cannons.

- Babur's horse archers were in left, right, and center formations, and Babur himself was with the center group. He secured one flank of his line against the structures of the town of Panipat, and the other against the banks of a river. The left and right cavalry wings were to circle around the flanks of the enemy, shower them with arrows from the powerful Mongol bows, and herd them toward the middle, impeding their mobility by crowding them and providing a dense target for the gunpowder weapons.

- On April 21, 1526, the sultan commanded his men to advance. His strategy was unimaginative, consisting of a frontal charge and the hope that his greater numbers would overwhelm the enemy. The charge seems to have been badly coordinated, and it bogged down among the carts and obstacles.

- Meanwhile, Babur's cavalry were deploying on both sides as planned, hemming in and harassing the attacking ranks. An attempt to break Babur's line near the town was repulsed, and surrounded and under heavy fire from all directions, the sultan's men were cut down in large numbers. Lodi led a final charge of the remaining 5,000 men in his reserve and, somewhere in the melee, was killed.

- Babur had defeated a much larger force by virtue of superior generalship, training, and technology. He offered a harsh but probably accurate assessment of his opponent: "Ibrahim Lodi was an inexperienced man, negligent in his movements. He marched without orders or halted without plan and engaged in battle without foresight."

Effects of the First Battle

- Babur was succeeded by his son, Humayan, who died in 1556. Humayan's son and heir, Akbar, was 13, and the Mughal dynasty seemed on the verge of coming to a premature end. Several strong local rulers who resented Mughal control took advantage of Humayan's death and Akbar's youth to rebel.

- The nominal leader of this rebellion was Sultan Adil Shah Suri, but the driving force behind its success was the sultan's prime minister, a Hindu named Hemu. Under his leadership, Delhi itself was captured, and the young Akbar was reduced to a refugee.

- Akbar's advisors counseled retreating to the traditional stronghold of Kabul and conceding the loss of India, but in this crisis, Akbar revealed himself to be a youth in the mold of the young Babur. Supported by Bairam Khan, an experienced Mughal general, Akbar decided to march south immediately and challenge Hemu for control over Delhi.

The Second Battle of Panipat (1556)

- Hemu decided to face Akbar at Panipat and dispatched his advance guard and artillery well ahead of his main army. This decision turned out to be a major mistake, because Akbar's own advance forces detected the inadequately guarded cannons, and the commander of Akbar's advance force immediately launched an attack.

- This sudden strike caught Hemu's men by surprise, and they ran, abandoning the guns. The result was that Akbar captured Hemu's entire artillery park intact and used it to augment his own cannons. Thus, in the coming battle, Hemu had no cannons, while Akbar had many.

- Despite this setback, Hemu moved with his main army to Panipat, and on November 5, 1556, the battle was fought. The true strength of Hemu's army was around 30,000 skilled Rajput and Afghan horsemen and 500 to 1,000 war elephants, many of which were encased in heavy plate armor.

- The smaller Mughal army had around 10,000 to 15,000 excellent horse archers, supplemented by some infantry and the matchlock men and artillerymen. The true commander on the Mughal side was the experienced general, Bairam Khan. Akbar was present, though he seems to have been stationed well to the rear and probably played little role in the actual oversight of the battle.

- Both armies were deployed in five main sections: a left wing, a right wing, a larger center block, a substantial advance force, and a reserve, all fronted by war elephants, with their leaders in the center.

- Hemu's right and left wings ferociously attacked their Mughal counterparts, led by the heavily armored elephants and backed up by elite cavalrymen. Both the Mughal wings began to give ground before the vicious onslaught, but the Mughal soldiers did not panic and maintained their formations.

- Meanwhile, elements of the Mughal cavalry on the extreme right and left sides rode in wide, sweeping arcs around the edges of the battlefield to attack Hemu's army from the sides and rear raining flights of arrows into the enemy ranks. Hemu rallied the forces immediately around him and organized a series of counterattacks, which succeeded in driving off the harassing Mughal cavalry columns.

- He then turned his attention to a renewed frontal assault against the Mughals. Some sources claim that this attack was on the verge of breaking the Mughal formation, which almost certainly would have led to victory, when an arrow found the weak spot of Hemu's helmet and pierced his eye. Seeing their general fall, his soldiers lost heart, the charge fizzled, and Akbar achieved victory.

Effects of the Second Battle

- The first victory of the Mughals at Panipat in 1526 opened the door for their advance into India, and the second, 30 years later, solidified their control.

- Akbar ruled for a half-century, becoming the greatest of the Mughal emperors. He and his armies conquered much of the Indian subcontinent and established a dynasty that continued until the arrival of the British several centuries later.

- Akbar was an energetic ruler who profoundly changed the culture and institutions of India. He patronized the arts, established the Mughal capital at Agra, regularized the currency, reformed the nobility, reorganized the military, restructured taxation, and to some degree, transitioned the Mongol steppe raiders into a more sedentary lifestyle. Despite these accomplishments, however, it should not be forgotten that the Mughals were, in essence, invaders, and their conquests were accompanied by copious bloodshed.

- By the time Akbar died in 1605, Mughal rule was firmly established—the true legacy of the battles of Panipat.

Suggested Reading

Babur, *The Baburnama*.

Sandhu, *A Military History of Medieval India*.

Singh, Harjett, *Cannons versus Elephants*.

Verma and Verma, *Decisive Battles of India through the Ages*.

Wink, *Akbar*.

Questions to Consider

1. How did Babur's early experiences shape or affect his later successes?

2. In the long run, do you think the First or the Second Battle of Panipat was more important, and why?

1571 Lepanto—Last Gasp of the Galleys

Lecture 18

By the 16th century, an annual ceremony held in the lagoon of Venice had become a spectacular maritime festival. The focal point of the ritual was the *Bucentaur*, the royal state galley of the Venetian Republic. Once this ship was launched, the doge of Venice held aloft a golden ring and solemnly pronounced the union of the city and the sea. This odd wedding ceremony symbolized the association between the prosperity of the Venetian Republic and its control over the Mediterranean. In earlier times, the ceremony would have affirmed the maritime strength of Venice, but by the mid-16th century, the ritual had become tinged with apprehension brought on by a new and terrifying Islamic power: the Ottoman Turks.

The Opponents

- The Turks had burst out of Anatolia, toppling the once-mighty Byzantine Empire and capturing the great city of Constantinople in 1453. Led by contingents of elite Janissary warriors, Ottoman Turkish armies had beaten Persia, overrun the entire eastern Mediterranean, and then moved onto the sea, snapping up island after island and fortress after fortress. Dozens of Venetian outposts in the eastern Mediterranean were pillaged or captured.

- The remaining major outpost of Christendom in the eastern Mediterranean was the island of Cyprus, held by the Venetians, and in 1570, the Ottomans turned their attention to it. The threat to Cyprus finally made the feuding Christian kingdoms of the western Mediterranean band together to try to stop the Turks.

- The Battle of Lepanto, the largest naval battle of the Renaissance, was a cataclysmic struggle that pitted the seemingly unstoppable Ottoman Turks against a desperate Christian naval coalition that included the Venetians, the pope, the Knights of Malta, and Spain.

- Not only would Lepanto prove to be a turning point in Ottoman naval expansion, but it constituted the final chapter of a type of naval warfare that had remained remarkably static for the first 3,500 years of recorded human history: the clash of oared war galleys.

The Ships

- The dominant warship that had been in use was a long, slender wooden galley propelled by hundreds of oarsmen, and the tactics still included ramming and boarding enemy vessels. Gunpowder weapons had begun to make an appearance, and each Renaissance galley was equipped with a large cannon at the bow and a few smaller ones to either side. Some of the soldiers manning the deck of the galley were also equipped with fairly primitive handheld firearms.

- The Turks had been slower to incorporate these new weapons into their ships; thus, Turkish galleys tended to have somewhat smaller and fewer cannons. Similarly, their crews made greater use of the traditional missile weapons rather than the new guns. Although Turkish ships were typically smaller than their Christian counterparts, because they lacked the weight of heavy cannons, they were also faster and more maneuverable.

- The Christian fleet included six Venetian galleasses, a bizarre and flawed hybrid of old and new technologies. These ships were broader and heavier than war galleys and had wooden castle-like structures at their front ends. Up to nine heavy cannons were placed in these castles, and lighter ones were added along the sides and at the stern, uniquely allowing the galleasses to fire in all directions.

The Campaign

- Pope Pius V had brought together the Christian coalition that fought at Lepanto. Realizing that no single Christian power could stand up to the Turks, Pius made it his personal project to create a Christian alliance that would be strong enough to do so.

- Accordingly, the Holy League was formed, consisting of the Papal States, Spain, Venice, Genoa, Tuscany, Naples, Sicily, and the Knights of St. John in Malta. Collectively, they mustered a fleet of approximately 200 ships, 30,000 soldiers, and 40,000 oarsmen.

- Command of the coalition settled on Don Juan of Austria. The fleets donated to the joint effort all had their own commanders, and there was tension among them, especially between the Spanish and the Venetians.

- In overall charge of the Turkish fleet was Ali Pasha, commanding about 200 galleys and more than 100 lighter combat vessels manned by roughly 30,000 soldiers and 50,000 oarsmen. Thus, the scale of the Battle of Lepanto was enormous, involving approximately 140,000 men on board nearly 600 ships.

- Although the Holy League had been formed to defend Cyprus, assembling the forces took far too long, and the Turks were able to conquer most of the island while the Christian fleet slowly made its

The only fortress holding out on Cyprus while the Christians assembled their forces was Famagusta, but after a bitter siege, this, too, fell to the Turks.

way across the Mediterranean to the harbor of Lepanto in the Gulf of Corinth.

- Strictly speaking, the loss of Cyprus also removed the urgency to fight the Turkish fleet, but word of the cruel treatment that had been inflicted on the Christian officer in charge so inflamed the Venetians that they demanded immediate battle. The fleets met on October 7, 1571, just off Scropha Point in the Gulf of Patras.

The Battle

- The Holy League deployed its ships into four squadrons:
  - On the left wing were 53 galleys under the command of Agostino Barbarigo.
  - In the center were 52 galleys led by Don Juan in his flagship, the *Real*.
  - On the right were 53 galleys under the command of Gian Andrea Doria.
  - Behind this line was a rearguard of 38 galleys under Don Alvaro de Bazan.
- Each of the three main squadrons had two galleasses, which were towed to a position about 500 yards in front of the Christian battle line. The galleasses of the northern and central divisions made it to their appointed stations, but the pair assigned to the south were held up by contrary winds and their own unwieldiness and lagged to the rear.
- The Turks mirrored this formation: a center squadron of 87 galleys under Ali Pasha in his flagship, the *Sultana*; a northern wing of 60; a southern one of 61 under the capable admiral Uluch Ali; and a small rearguard of 8. As was customary, the Turkish formation adopted a shallow crescent shape, while the Christians maintained a straight battle line.

- The subsequent battle began in the north when the two galleasses stationed there opened fire on the oncoming Turkish galleys. A lucky shot holed one of the larger Turkish ships beneath the waterline near the bow, and it began to sink. As the Turkish galleys, eager to get to grips with the main line of the Holy League, sped past the galleasses, they took a heavy beating from the Christians' guns, disrupting the Turkish formation.

- Meanwhile, in the center, the galleasses had opened fire with similar results, and the two lines converged. A long hand-to-hand struggle followed, as ships ground against one another and their crews swept back and forth, alternately boarding other vessels and being boarded themselves.

- The *Sultana* was mobbed by coalition ships. After hours of fighting and three separate attempts, the *Sultana* was boarded, its crew overcome and slaughtered, and Ali Pasha himself slain. The fighting in the center continued for another half hour or so, but the Christian ships now had the advantage, and large numbers of Turkish galleys began to surrender.

- In the south, the Turkish commander Uluch Ali and the Christian leader Andrea Doria both eschewed a head-on charge in favor of trying to outflank each other. The net effect was that both southern squadrons began to angle away further to the south, creating a gap between the southern squadrons and the rest of the battle.

- Seeing that he could not get around his foe by moving south and that some of the Christian ships had become detached from the main group, Uluch Ali turned his ships north and drove for the gap.

- Had Ali charged back to the center a bit earlier, he could have had a great effect on the overall battle, but the fight in the center had already been won by the Holy League, and Christian ships now began to converge on him from all sides. By midafternoon, the battle was over.

Outcomes

- For the Christians, Lepanto was a stunning victory, made all the more glorious by being unexpected. On the other side, the Ottomans consoled themselves by ascribing the defeat to the inscrutable will of God and asserting that, when the entire campaign was considered, even though they had lost a fleet, they had gained a more valuable prize in Cyprus.

- Lepanto might have led to substantial gains for the Christians, but their always-fractious coalition quickly fell prey to rivalries and infighting. They thus failed to exploit the opportunity that the victory of Lepanto offered them. The Mediterranean region settled into a standoff, with each side retreating into its own domain.

- The Turks quickly rebuilt a sizable fleet, but their maritime ambitions were severely and permanently checked. They could replace ships fairly easily, but they could not replace their trained crews so readily. Lepanto marked the end of major Turkish raids on the western Mediterranean and the effective end of their assaults on the key island and port outposts of Christendom.

- Although Ottoman ambition may have been checked on the water, it was by no means extinguished. The Ottoman Turks would simply seek other directions in which to extend their influence and power, and their eyes soon focused on their land border with central Europe as the stage for their next great thrust into Christendom. The final showdown between the two forces would take place a century later, on the plains of central Europe, with the great siege of Vienna in 1683.

Suggested Reading

Beeching, *The Galleys at Lepanto*.

Bicheno, *Crescent and Cross*.

Crowly, *Empires of the Sea*.

Questions to Consider

1. In what ways did changing technology affect the outcome of the Battle of Lepanto?

2. In what ways did leadership and decision making affect the outcome of the battle?

1592 Sacheon—Yi's Mighty Turtle Ships
Lecture 19

In July of 1592, the Japanese commander who stood overlooking the Bay of Sacheon in Korea was probably quite pleased. The invasion of Korea was proceeding as planned; a force of more than 160,000 Japanese samurai had captured Pusan, Korea's main port, then marched north and seized the capital city of Seoul. The great Japanese warlord Toyotomi Hideyoshi, the mastermind behind the invasion, was well on his way to extending his dominion to include Korea and, using that foothold, to conquer China. The only blotch on the invasion's record of triumphs occurred when an unusually aggressive Korean admiral, Yi Sun-shin had caught and destroyed several dozen Japanese vessels. Now, this same Yi Sun-shin had been sighted sailing into Sacheon Bay.

The Battle of Sacheon

- On July 8, 1592, the Korean admiral Yi Sun-shin had been sighted sailing into Sacheon Bay at the head of his squadron, and the commander of the Japanese invasion at once ordered his men to attack. Confirming the Japanese opinion of the unwarlike nature of the Koreans, as soon as the Japanese warships began pouring out into the bay, the Koreans retreated.

- The Japanese eagerly pursued, and the Korean craft suddenly turned smartly about and headed back toward the Japanese. The apparent flight had been a ruse to lure the Japanese out into open water. No matter; the Japanese were confident of success, and moved to engage the Koreans.

- Then a terrifying sight emerged: A dragon's head spitting smoke and flame connected to a strange, bulky, low-lying body. There was no open deck, and no humans were visible; instead, the creature had a curved, humped back formed from interlocking hexagonal plates with sharp, tapering spines. From small holes along its sides, cannons discharged a deadly hail. This monster plunged into the midst of the Japanese ships, spewing death and destruction.

- While the creature wreaked havoc and confusion among the Japanese, the rest of the Korean fleet, more conventionally designed vessels, stood at a distance and bombarded the Japanese with cannons and arrows.

- The Battle of Sacheon turned into a slaughter, with the confused Japanese trapped between the guns of the impregnable monster and the storm of missiles from the Korean ships that surrounded them.

- The battle marked the combat debut of the turtle ship. Over the next decade, Yi and his turtle ships would play a key role in defeating Japan's invasions of Korea, and Yi's actions would establish him as one of the greatest admirals of all time.

The Opponents

- Yi Sun-shin was born in 1545 and decided to pursue a military career. According to a story that may be apocryphal, the first time Yi attempted to pass the rigorous military entrance examinations, he fell off his horse, breaking his leg. Though he knew he had failed the exam, he splinted the break using the branch of a nearby tree and completed the exercise. After passing the exam in a later year, he rose through the ranks, distinguishing himself in action.

- Advancement in the Korean political and military structure was heavily based on favoritism and patronage rather than ability, and thanks to the support of a childhood friend who had influence at court, Yi was the commander of a small contingent of the Korean fleet when the Japanese attacked in 1592.

- Yi kept a diary that has mostly survived and is available in English translation. In addition to its historical importance for the firsthand descriptions it gives of major battles and key events surrounding the invasion of Korea, it also offers a unique and humanizing portrait of a legendary figure.

- At the Battle of Sacheon, one of the very few casualties was suffered by Admiral Yi, who was shot in the shoulder by a Japanese musket. This incident also proved prophetic, in that Yi would suffer several more combat injuries, culminating in his death in battle.

- The driving force behind the invasion was the Japanese warlord Toyotomi Hideyoshi. He joined the army of one of Japan's lords as a common foot soldier, repeatedly distinguished himself in battle, and rose to the level of general.

- Some opportune deaths combined with his own talent and ambition allowed Hideyoshi to become one of the major powers in Japan, and he solidified his reputation with a bold invasion of the southern island of Japan, Kyoshu. The conquest of Kyoshu had involved a large-scale amphibious landing, and this may have given Hideyoshi the idea for the invasion of Korea.

- Kyoshu is separated from the Korean mainland by only 30 miles of water, known as the Tsushima straits. Several islands in the straits provide useful way points and visual guidance. It is an easy crossing—not only an important trade route but an invasion path both to and from Japan.

Japanese Naval Warfare

- For the Japanese, battles at sea were simply extensions of land combat, and they attempted to use similar tactics. Thus, Japanese warships were designed less as weapons in their own right than as platforms for infantry, and the main tactic in naval battles was to run one ship up alongside another and let the two crews fight it out hand-to-hand. Primitive cannon were in use but were rarely placed on ships, and the firepower of a Japanese warship came entirely from the personal weapons carried by her crew.

- The samurai had a long tradition of excellence at archery, and every samurai would have been trained in the use of the Japanese longbow. Recently, however, gunpowder had arrived on Japanese battlefields, and a major component of armies was now foot soldiers armed with an early form of musket.

- The largest and most specialized type of Japanese warship, *ataka bune*, had high, flat, wooden sides pierced at regular intervals with loopholes through which the men within could fire their guns. A line of oars projected from a lower row of holes, and there was usually one mast with a square sail, typically lowered during battle. On the flat roof of the main cabin, more infantry were stationed and sometimes a tower. The crew consisted of around 80 oarsmen and 60 fighting men. As a ship, it was slow, ponderous, and not very seaworthy.

- Next in size was the *seki bune*, basically a smaller version of the *ataka bune*. It had a tapered bow and was either partially enclosed or had a waist-high railing protecting the crew. Somewhat more maneuverable, it carried around 40 oarsmen and 30 soldiers.

- The smallest, swiftest craft was the *kobaya*, which had an open deck and was really just a small ship carrying some soldiers. To offer some token protection, the Japanese often built a wooden framework from which were hung fabric screens. These might offer enough resistance to catch an arrow that was nearly spent, but would not have stopped a more forceful projectile.

Korean Naval Warfare

- The Koreans approached naval combat with a different strategy. Rather than closing with the enemy and boarding, they preferred to stand off at a distance and bombard an enemy vessel with cannonballs, fire bombs, bullets, and arrows.

- The standard Korean warship, the *panokson*, was medium-sized and had two decks: an enclosed lower one that protected the rowers and a wide upper fighting deck surrounded by railings. The larger models might also have a tower on the upper deck. They were solidly constructed to bear the weight of cannons.

- Each *panokson* had a variety of cannons of different sizes and ranges. All fired stone and iron cannonballs, as well as incendiary bombs. The Koreans also used them to propel enormous wooden arrows equipped with iron tips and leather vanes, said to cause massive destruction when they struck their target.

- Admiral Yi's famed turtle ships were propelled by the usual mixture of oars and a mast carrying a square sail, but because of their shape, they were especially maneuverable. They mounted about six cannon on each side, with several more firing to the front and back.

© Brücke-Osteuropa/Wikimedia Commons/Public Domain.

The turtle ships designed by Admiral Yi would play a key role in defeating Japan's invasions of Korea, and Yi's actions would establish him as one of the greatest admirals of all time.

- Because of its especially sturdy design and thick protection, the turtle ship was capable of ramming its opponents, but the preferred method of combat was to sink enemy vessels with shots from its cannons or to set them afire.

The Battle of Hansan Island

- Recognizing the danger Yi posed, Hideyoshi ordered his generals to destroy Yi and his ships, leading to Yi's greatest victory, the Battle of Hansan Island.

- Yi again used a false retreat by to lure the Japanese into a stretch of open water near Hansan Island. The Japanese raced after the Korean ships and found themselves confronting the main Korean fleet arranged in a crescent-shaped formation. The Japanese were drawn toward the center of this formation, where Yi encircled them.

- At the center of the Korean formation were the turtle ships, and Yi now ordered these to move forward to engage and hold the Japanese, while the *panokson* circled around them and began their bombardment. Once again, the Japanese were slaughtered, and the entire fleet was wiped out.

Outcomes

- Korea was weakened and devastated by the Japanese invasion, but the spirited defense offered by the Koreans eventually became a foundation of nationalism and independence.

- In Japan, the thwarted invasions marked the country's first attempt to acquire an empire on the mainland of Asia—a dark ambition revived with global effects in the 20th century.

Suggested Reading

Park, *Yi-Sun Shin and His Turtleboat Armada.*

Turnbull, *The Samurai Invasion of Korea.*

———, *Fighting Ships of the Far East*, vol. 2.

Yi, *War Diary of Admiral Yi Sun-Sin.*

Questions to Consider

1. How do the various ship designs described in this lecture highlight different weapons and styles of fighting?

2. How do you evaluate Admiral Yi's qualities as a commander compared to other "great" leaders, such as Ramesses, Alexander, Genghis Khan, or Babur?

1600 Sekigahara—Samurai Showdown

Lecture 20

On the eve of a civil war in Japan, two friends spent a convivial evening together. They were Tokugawa Ieyasu, the most powerful lord in Japan, and Torii Mototada, the commander of one of Tokugawa's castles. The next morning, the friends parted, knowing they would never meet again. Both understood that as soon as war broke out, their enemies would attack the castle; Mototada and the defenders would be outnumbered, yet as samurai, they would willingly stay and die for their lord. Soon after, 40,000 warriors besieged the castle, but Mototada held them off for 10 days, giving Tokugawa time to muster his own armies. Tokugawa went on to confront his enemies at the Battle of Sekigahara, the most decisive battle of the samurai era.

The Opponents

- The Battle of Sekigahara would turn on the personalities of three people, all of whom had ties to Toyotomi Hideyoshi, the great warlord who had unified Japan and launched the invasions of Korea that had been repelled by Admiral Yi and his turtle ships.

- The last of the unsuccessful Korean invasions had been dispatched in 1597 under the command of Toyotomi's adopted son, a young samurai named Kobayakawa Hideaki. He was a figurehead as commander; much of the real decision making was done by a council of generals whose squabbling was a major factor in the failure of the expedition.

- As the notional commander, Kobayakawa was summoned back to Japan in disgrace. Toyotomi apparently planned to punish him for his failure, but Tokugawa, who was already one of the most important lords in Japan, persuaded Toyotomi to be lenient.

- Tokugawa was an experienced and wily general who had fought many battles and was a skilled tactician and strategist. He had helped Toyotomi gain control of Japan while biding his time and building up his own strength. He was in his mid-50s and at the apex of his wealth and power, with a network of allies who were loyal friends. Yet in his climb to prominence, he had made a number of enemies, and there were many who resented his prosperity.

- In May 1598, near death, Toyotomi summoned the five most powerful lords, including Tokugawa, to act as a council of regents and rule the country in concert with five administrators until his old son was old enough to rule.

- The leader of this group was Ishida Mitsunari, a descendant of a famous family that had fallen on hard times. It was these three men: Tokugawa, the powerful general and wealthy lord; Ishida, the clever administrator; and Kobayakawa, the young samurai, who would determine the outcome at Sekigahara.

- Soon, two coalitions began to emerge: one centered on Tokugawa and his long-time friends and allies and the other surrounding Ishida. In an official sense, Kobayakawa was clearly part of Ishida's faction because Ishida was defending the legacy and wishes of Kobayakawa's adoptive father, Toyotomi. This was also the group that supported the claim of Kobayakawa's adoptive brother, Toyotomi's 5-year old son. On the other hand, Kobayakawa remembered the great favor that Tokugawa had done for him after the Korean debacle.

The Armies and Their Technology

- The armies at Sekigahara fought with nearly identical weapons and armor.
  - The elite samurai were highly trained warriors proficient with a variety of weapons, including sword and bow. Some were mounted, although many fought on foot.

  - The *ashigaru* were lower-ranked and carried long stabbing spears called *yari*, while others were trained to fire primitive matchlocks.

- Guns were a relatively new innovation, probably introduced by Portuguese traders in the mid-1500s, and the Japanese quickly learned to manufacture them. On the battlefield, the troops thus armed were deployed in mass blocks, where their volleys would have a shock effect.

Samurai were bound by a rigid code of behavior, sometimes called bushido—"the way of the warrior."

- Soldiers were equipped with a variety of body armor, ranging from solid breastplates to cuirasses formed of hundreds of metal plates held together with intricate bindings. This armor was frequently lacquered and, thus, could be brightly colored. Metal helmets were standard, as were shin and arm guards. Military units were not standardized; each lord organized his men as he wished.

The Battle

- The actual battle was preceded by several months of hostilities, during which each side attempted to seize key castles occupied by the other. Ishida's forces captured the castle of Fushimi, held by Tokugawa's friend Mototada. Other Ishida-faction armies took Tanabe Castle and Otsu Castle, but these local victories obviously occupied some of Ishida's forces so that they were not present at Sekigahara.

- Meanwhile, Tokugawa's generals seized the strategically located Gifu Castle, but Tokugawa's son allowed himself to be delayed unnecessarily in attempting to take Ueda Castle. All these moves were like a chess game, in which the opposing sides sometimes expended strength to gain a strategic location or deliberately sacrificed a piece in one place to weaken an opponent's attack in another.

- Much of the maneuvering concerned control of two key roads: the coastal Tokaido road and the inland Nakasendo road. There is a point at which the main Japanese island of Honshu narrows, and both roads must squeeze through a slender gap between Lake Biwa and the Pacific Ocean. Near this natural chokepoint, where many of the contested castles were located, the two armies finally met on October 16, 1600.

- Sekigahara lies in a constricted valley. Ishida's forces converged here on the night of October 15 and deployed on the hillsides and on bits of high ground, arrayed in a roughly semicircular shape. Ishida's plan was to force Tokugawa's army to cross the swampy lowlands and attack uphill against his fortified positions, where they would be surrounded on three sides.

- Ishida was stationed toward the left flank with about 8,000 men; from there, contingents under various lords stretched south in a curving line. Near the center was a block of 17,000 troops under the command of Ukita Hideie. On the far right flank, Kobayakawa was stationed at the head of 16,000 of his own men.

- Across the valley, Tokugawa positioned himself on a hillside behind a double line of units totaling around 40,000. He held back his own 30,000 personal retainers in a block beside him as a reserve.

- Both armies entered the valley during the night, but a thick mist obscured their movements so that neither was sure where the other was. As dawn broke, one of Tokugawa's lords sprang forward at the head of 30 of his mounted samurai and launched a charge against Ukita's center division. Following his lead, other Tokugawa units charged straight across the valley and engaged Ishida's center and left divisions.

- Ishida's men were managing to resist the assault, but as he looked along the line, he noticed that one important contingent under Lord Shimazu was not participating. When messengers brought no explanation, Ishida went personally to see Shimazu and was brusquely informed that Shimazu would enter the battle when the time was right.

- By late morning, most of Tokugawa's forces, except for his reserve, had been drawn across the field and were heavily engaged. This was the moment that Ishida had been waiting for, and he gave the signal for Kobayakawa to descend from the hillside and engage Tokugawa's army from the side. Kobayakawa did not respond. Even worse from Ishida's perspective, several minor lords on the right flank of his army now emulated Kobayakawa and held back their troops, as well.

- Across the valley, Tokugawa was keeping an eye on Kobayakawa's formation, well aware of the danger that Kobayakawa posed to the flank of his attacking forces. Like Ishida, he became focused on what the young lord would choose to do. Finally, Tokugawa determined to force Kobayakawa's decision; thus, he ordered a few of his musketeers to fire in the direction of Kobayakawa's men.

- The shots had an immediate effect. Kobayakawa sprang to his feet and exclaimed, "Our target is Otani!"—one of Ishida's commanders. Kobayakawa's men poured down the mountain and smashed into Otani's regiment. Kobayakawa had chosen to switch sides and join Tokugawa.

- This treachery did not come as a total surprise to Ishida and his men. They had been aware of Kobayakawa's long vacillation, and Otani in particular had been so suspicious of his supposed ally that he had ordered a section of his men to be ready to turn and face Kobayakawa if he switched sides. Thus, Kobayakawa's attack did not have the destructive impact it might have had against a wholly unprepared foe.

- Otani's men fought fiercely, but they were vastly outnumbered—a disadvantage that multiplied when the minor lords who had also been delaying now followed Kobayakawa's lead, switched sides, and threw their lot in with Tokugawa.

- At the other end of the line, Shimazu had finally joined the battle, but it was much too late, and the tide was now turning against Ishida. With Tokugawa's forces pressing in from the front and Kobayakawa's from the side, Ishida's lines began to buckle, and individual commanders and units started to break away. By early afternoon, the victory would be Tokugawa's.

Outcomes

- After Sekigahara, Tokugawa's power was unrivaled. He redistributed land to reward his lords and punish his enemies. The still preadolescent son of Toyotomi Hideyoshi, in whose name Ishida had fought, was granted a large territory and allowed to live in Osaka castle. In 1615, however, trumped-up charges were brought against him, the castle was besieged, and the last members of the Toyotomi clan were either killed or committed suicide.

- In 1603, Tokugawa took the title of shogun and effectively became the ruler of a united Japan. In 1605, he officially retired, although he continued to manipulate events behind the scenes. The dynasty that he founded, the Tokugawa shogunate, continued to rule Japan for approximately 250 years.

Suggested Reading

Bryant, *Sekigahara 1600.*

Sansom, *A History of Japan, 1334–1615.*

Turnbull, *Battles of the Samurai.*

Questions to Consider

1. What aspects of samurai warfare reflect distinctive qualities in Japanese culture at this time?

2. Which side do you think Kobayakawa should have chosen to align himself with and why?

1683 Vienna—The Great Ottoman Siege

Lecture 21

On August 6, 1682, outside one of the gates to Topkapi Palace in Istanbul, seven standards were planted in the earth, indicating that the sultan was poised to embark on a campaign. This sultan was Mehmed IV of the mighty Ottoman Empire, and the military expedition thus begun was one of the most significant in Ottoman history. Its goal was to capture Vienna, a city that had stood for centuries as a fortress protecting the eastern flank of Europe from Ottoman expansion. Vienna had repulsed one great Ottoman onslaught in 1528. Mehmed was determined to end these centuries of defiance, to place his name among the legendary Ottoman conquerors of the past, and to bring Vienna to its knees.

The Opponents

- Clearly a primary motive for the campaign to take Vienna was Mehmed IV's ambition to go down in history as one of the great conquering sultans. Early in his reign, he had achieved successes both in the Mediterranean and in Europe, but he needed one spectacular military victory to cement his reputation.

- Mehmed's grand vizier at the time of the attack on Vienna was Kara Mustafa Pasha. His origins are somewhat mysterious. By virtue of either talent or personality, he rose quickly from messenger to military commander to minor vizier and, eventually, was appointed commander of the imperial fleet. In this role, he directed the conquest of several Aegean islands, leading to his appointment as deputy grand vizier and, finally, grand vizier.

- The Ottoman Turkish army had long enjoyed a reputation for innovation and excellence, but its tactics were beginning to stagnate.
    - In keeping with its steppe traditions, it still included a large contingent of heavily armored *sipahi* horsemen, who could fight with a variety of weapons.

    - These were augmented by units of Tatar light cavalry, who still fought, like their Mongol ancestors, as horse archers practicing hit-and-run tactics.

    - The infantry included engineers and artillerymen, as well as foot soldiers.

- The elite troops of the Ottoman Turks remained the Janissaries, trained like the Mamluks in military academies, where they were converted to Islam and raised to be fanatically loyal and well-trained soldiers. They were highly flexible troops, able to fight with sword, bow, or gun.

- One weakness of the Ottoman army was its artillery. Although the Turks possessed large cannon well-suited to battering down fortifications, heavy rains had made hauling them along the roads to Vienna logistically impracticable. Thus, they had to subdue Vienna using medium-size cannon firing 24-pound balls.

- The Turks hoped to make up for this defect by the excellence of their sapper corps, which tunneled beneath the walls and set off mines to demolish them. The Turks were particularly skilled at this sort of warfare, and the attacking army contained large numbers of experienced miners and engineers.

- Facing this threat was Leopold I, the ruler of Austria and the Holy Roman Emperor. He was a member of the Hapsburg family, which controlled many countries; Leopold knew multiple languages and was an avid reader and student of history. Probably his greatest defect as a leader was vacillation.

The Campaign

- The Ottoman army that set out for Vienna in October 1682 probably consisted of around 100,000 men. By the end of June 1683, the Turkish force moved into enemy territory and bore down on Vienna.

- As the Turkish army steadily advanced, Leopold and his advisors dithered over what the Turks' objective was, thus preventing any decisive steps to defend Vienna. When Leopold was indecisive, he relied heavily on prayer; during the crucial last month before the Turks arrived, rather than fortifying the defenses or bringing in reinforcements, Leopold ordered that prayers be recited in St. Stephen's Cathedral 24 hours a day and that all members of the city's guilds attend.

- Leopold not only attended services but continued his normal routine, even going hunting outside the city as late as July 6. The next day, new reports testifying to the rapid approach of the Turks, confirmed by hordes of refugees fleeing in front of them, finally broke through. Suddenly acknowledging the reality and proximity of the threat, Leopold abruptly announced that the royal family and court would flee that very night.

- Leopold left so quickly that the army officer he had appointed to defend Vienna, Rudiger von Starhemberg, was still in the field and had to be summoned. Von Starhemberg faced a daunting task: The Turks would arrive in six days, the garrison numbered only about 2,000 men, the fortifications were in a state of neglect, and the city had not been fully provisioned for a siege.

- By the time the Turks arrived on July 14, Vienna's garrison numbered about 15,000 men, and more than 300 cannons of various sizes were available to the defenders.

- The defenders' greatest concern was the city walls.

Another crucial weapon in good supply was hand grenades—80,000 had been stockpiled, and nearly all would be expended during the siege.

- The arrival of large cannon on the battlefield had rendered obsolete the high stone walls of medieval fortresses, which proved too brittle to resist the smashing effects of cannonballs. As a result, fortifications, such as Vienna's, had been rebuilt to feature extremely thick, but relatively low earthen embankments that could absorb a cannonball's impact.

- Also, the straight walls of medieval castles had been replaced with star-shaped fortresses whose sides were studded with projecting triangular bastions. They also commonly featured detached triangular outworks, called ravelins, situated beyond the main walls. All these were built on along carefully calculated geometric lines so that shots fired from the cannons and muskets mounted on the projecting bastions and ravelins created interlocking fields of fire, with the gunners in one section able to blast away any attackers threatening a neighboring section.

- But Vienna's fortifications did not conform to this mathematical ideal. Some stretches of the wall were soundly designed, but others constituted weak spots certain to draw the attention of a knowledgeable attacker. Unfortunately for the defenders of Vienna, the Turks were experienced at siege warfare.

The Siege

- Arriving at Vienna on July 14, Kara Mustafa erected his tent opposite the most vulnerable point. The defense of this section depended on two bastions projecting from the walls and an outwork just in front of them. The defenders knew that this was their weak spot, and roughly half the garrison was assigned to defend it.

- Kara Mustafa sent an envoy to the defenders bearing a demand for surrender. It was rejected, and the siege began. About 20,000 of the Turks were directly engaged in the siege, while the other 70,000 cut off the city from outside aid and ravaged the countryside. Meanwhile, the Tatar light cavalry engaged in long-range raids, traveling hundreds of miles into enemy territory.

- Within 24 hours, the Turkish engineers had dug protected positions for their cannons, made an observation bunker for Kara Mustafa overlooking the key zone, and begun a series of trenches toward the walls. By July 22, the trenches had advanced within a few feet of the defenses, and the first serious assaults began.

- The Turkish miners had also burrowed a series of tunnels. The first mine was set off on July 23, followed by a larger one on July 25 that blew up part of a bastion. Janissaries poured into the breach but were slaughtered by heavy gunfire and grenades.

- On August 12, the Turks exploded two especially large mines, blasting a clear pathway into the outwork. Despite desperate efforts by the defenders, the Janissaries flooded into the gap and, after a battle lasting several hours, succeeded in capturing the ravelin. The first key strongpoint had fallen.

- Von Starhemberg took an active role in the defense, personally leading a countercharge against some Turks who had gained a foothold on the walls and organizing sorties to harass the Turkish digging parties. On August 25, one of these was so successful that it managed to briefly capture one of the main Turkish batteries of cannons. The Viennese could not disable the guns, however, and retreated, having suffered 200 casualties

- On August 27, von Starhemberg ordered that rockets be discharged from the spire of St. Stephen's Cathedral to signal that the garrison could not hold out much longer. By September 8, both the bastions were so heavily damaged that their effectiveness was greatly reduced, and the number of fit soldiers to man the defenses was down to 4,000.

- While Vienna stubbornly resisted, Leopold had formed an alliance with Poland-Lithuania and had been assembling a relief army led by King Jan Sobieski and Charles, duke of Lorraine. They had amassed a force of around 60,000 that was especially strong in heavy cavalry.

- The Turks had neglected to establish strong defenses against an outside force; thus, Sobieski and Charles were able to infiltrate through the Wienerwald to the outskirts of Vienna before dawn on September 12. Kara Mustafa hurriedly assigned about half his army to meet this new threat, orienting them into a battle line.

- The result became known as the Battle of Kahlenberg. It unfolded in a confused series of encounters as the two sides advanced along a four-mile front over rough terrain. By late afternoon, the allied army had nosed forward onto the plains leading to Vienna, and both Charles on the left and Sobieski on the right sensed that the crucial moment had arrived. Both ordered their men to the attack.

- Although the Turks resisted bravely, this charge proved decisive, and the Ottoman lines disintegrated. Kara Mustafa escaped, taking with him what was left of the army's treasury, but most of his army was destroyed. In the campaigns that immediately followed, the Ottomans lost a chunk of their European territory, including, most importantly, Hungary.

- Much of the previous 200 years of European history had been dominated by Turkish attempts to expand their reach into Western Europe, resulting in a series of epic sieges and battles. Vienna decisively marked the end of these attempts and was the turning point in Ottoman history.

Suggested Reading

Millar, *Vienna 1683*.

Stoye, *The Siege of Vienna*.

Wheatcroft, *The Enemy at the Gate*.

Questions to Consider

1. What factors that are important in siege warfare do not play as great a role in battles between two armies in the field?

2. What should the Ottoman Turks have done differently that might have resulted in a victory at Vienna?

1709 Poltava—Sweden's Fall, Russia's Rise

Lecture 22

The Battle of Poltava in 1709 marked a turning point in military history—the moment when the dominant force on the battlefield shifted from edged weapons driven by human muscle, such as swords, spears, and arrows, to cannons and guns using gunpowder to propel bullets, shells, or cannonballs. One of the main agents for this transformation came from a perhaps unexpected quarter: the rather obscure kingdom of Sweden.

Military Innovations in Sweden

- In 1611, 17-year-old Gustavus Adolphus ascended to the throne of Sweden. He proved to be a far-sighted military genius, radically restructuring the Swedish army around the principles of firepower and mobility. He introduced a new, much lighter musket and reversed the percentage of his infantry who carried guns versus those who wielded pikes, for the first time making guns the primary weapon of an army.

- Gustavus equipped the Swedish army with three standard-size cannons, emphasizing light cannons that could be drawn by a few horses and deployed in accordance with troop movements on the battlefield. Cavalry were used as shock troops, and soldiers were organized into regiments.

- Using his new, modern army, Gustavus embarked on a series of spectacularly successful campaigns and turned Sweden into the most powerful state in northern Europe. By the late 1600s, Sweden dominated Scandinavia and the Baltic Sea and had acquired footholds in Germany that allowed it to control commerce on three of Germany's major rivers.

The Opponents

- In 1697, 15-year-old Charles XII came to the throne. Sweden's chief victims over the previous century—Denmark, Poland, and Russia—

sensed an opportunity to take advantage of the inexperience and youth of the new monarch. The result was the Great Northern War: Denmark invaded Schleswig-Holstein, Poland attacked Riga, and the Russians invaded Livonia.

- Unfortunately for them, Charles XII turned out to be a military leader fully as talented as his ancestors. Now 18, he personally led the Swedish army in a counterattack, first defeating Denmark, then the Russians, and finally embarking on a brilliant campaign in Poland that lasted several years and produced at least six major victories.

- Charles XII's personal role model was Alexander the Great. Well aware that Alexander had conquered a vast eastern empire with a relatively small army, Charles now determined to invade Russia and capture Moscow. The army with which Charles proposed to do this numbered around 40,000. On September 7, 1707, Charles and his forces crossed the Oder River into Poland.

- His opponent was another young, determined ruler: Peter I of Russia, who came to the throne determined to change Russia's feudal economic and social structure. He was fascinated by technology and had traveled extensively in Western Europe, eagerly absorbing the most sophisticated cultural and scientific achievements of the age.

- One of Peter's desires was to make Russia a significant

Peter the Great is known for creating the first Russian navy; he even worked incognito in an English shipyard to learn techniques for constructing warships.

naval power, but he needed access to either the Baltic or the Crimean Sea, preferably both. The Swedes blocked his way in the north and the Turks, in the south.

- At the time of Charles's invasion, Peter had begun his modernization of the Russian army, probably most successful in the area of upgrading the army's gunpowder weapons, from the common soldier's musket up to heavy siege cannons. The Russian army was especially well-equipped with a large number of modern artillery pieces.

The Campaign

- Peter adopted a strategy emulated during later invasions of Russia by Napoleon and Hitler. His army retreated before the advancing Swedes and enacted a policy of destroying farms, burning crops, and forcing the invaders to maintain long supply lines. Peter was also counting on one of Russia's most potent defensive strengths: its long, terrible winters.

- Peter's other secret weapon was the vast size of Russia: He could afford to surrender huge chunks of territory, luring the Swedes farther from home, reinforcements, and supplies.

- The Swedish army crossed the Vistula River in December 1707 and steadily progressed across Poland and into Lithuania before halting for the winter in February. The army moved forward again in June, and there were several skirmishes and minor battles with the Russians. Peter and his main army shadowed the Swedes but would not commit to a decisive battle, thus drawing Charles deeper into Russia.

- The supply question was becoming critical, and Charles's best bet seemed to be to turn south toward the Ukraine, although this would divert his march away from Moscow. When some Ukraine-based Cossack groups rebelled against Peter and offered to join Charles, also offering to feed his starving army, his decision to turn south was reinforced.

- The winter of 1708–1709 turned out to be one of the coldest and harshest in memory. Charles's army plodded forward, continuing the campaign through the winter. After several skirmishes with the Russians, it came to a stop on the main road from Kiev to Kharkov at a point between the Vorskla and Psiol rivers.

- The Russians were encamped nearby, outside the small town of Poltava. Disease, combat, and the winter had whittled Charles's army down to an effective fighting force of about 25,000 men. Not only was food running short, but the remaining store of gunpowder was limited and of questionable quality.

- On June 17, Charles was inspecting his outposts along the Vorskla River when he was struck in the foot by a Russian musket ball. He was used to leading from the front in battle, personally directing charges and rousing his men with his example, but this injury was severe enough that he had to be carried in a litter at the Battle of Poltava, both limiting his mobility and depriving his men of his crucial inspiration.

The Battle

- Peter entrenched his main army, about 25,000 infantry and 73 cannons, in a fortified camp backed against the Vorskla River, a few miles north of Poltava. To get at the camp, the Swedish army would have to pass through a gap between two dense woods to the southwest. Across this gap, Peter had erected a row of six redoubts: miniature rectangular or triangular earthen forts spaced at 150-meter intervals. He then built four more redoubts at a right angle to the first set.

- Together, the redoubts formed a T and were manned by about 5,000 soldiers with 16 cannons. To attack the main Russian force, the Swedish army had to pass these small forts, which could pour fire into their vulnerable flanks. Peter positioned the majority of his cavalry, about 9,000 men, in a line behind the redoubts.

- Charles's plan for the battle seems daring, even foolhardy. His available resources were about 8,000 infantry, 7,000 cavalry, and four small cannons. He proposed to rush this small force past the lines of redoubts before dawn, then assault the main Russian camp.

- Things began to go wrong almost at once. The infantry was supposed to lead, with the cavalry following close behind, but the two groups lost contact, and the infantry had to pause until the cavalry was located and brought forward, delaying the attack.

- Worse, the Swedes were detected by the redoubts, which opened fire at point-blank range. The columns of Swedish infantry furthest away were able to stay focused and hurry past, but those closer to the redoubts became bogged down as the men returned fire.

- Had the Swedish units continued moving forward after subduing the redoubts, they might have prevailed. But a group of six battalions, one-third of the entire Swedish infantry, under the command of General Roos, stopped altogether and became enmeshed in a brutal fight for control of the third redoubt.

- Meanwhile, the remaining infantry and the cavalry forged past the redoubts onto the plain in front of the Russian camp. Charles's forces had been badly mauled in passing the redoubts, but it seemed that they could proceed with the plan. Charles began to assemble and organize his forces in preparation for an assault on the main Russian camp. But he now realized that a third of his infantry was missing.

- Roos had stubbornly continued to assault the third redoubt in a series of futile attacks. Having lost 40 percent of his men, he withdrew to a nearby wood, where the Russians pounced. After a running fight, Roos was isolated from the rest of the army and down to 20 percent of his original men. He surrendered.

- Back out on the plain, every minute the Swedes delayed, they lost men to Russian cannon fire. Then, the Russians marched out and began to deploy in a line facing them. Now 22,000 fresh Russian troops began to march forward against the thin line of tired Swedish infantry, who numbered no more than 4,000.

- The two lines began about 800 meters apart. As the range decreased to 500 meters, the massed Russian cannon opened fire, smashing gaps in the Swedish line. At 200 meters, the Russian cannons switched from solid balls to grapeshot and scrap metal that spewed a storm of lethal fragments into the Swedish ranks.

- Finally, at 50 meters, the Russian infantry, four ranks deep, lowered their muskets and unleashed a colossal volley. All this time, the Swedish soldiers had marched forward without firing a single shot in return. Now, at a range of 30 meters, the survivors finally paused, discharged their muskets in one volley, and charged.

- Astonishingly, on the Swedish right, the thin, ragged line of infantry actually began to force back the thicker Russian formation. On the left, however, the line began to falter. The longer Russian line now closed around the Swedes, trapping and attacking them from three sides. Under this pressure, the Swedish charge faltered, and the battle was transformed into a slaughter.

Outcomes

- The Battle of Poltava marked the end of the Swedish Empire and of Sweden's role as a great military power. The state that had looked poised to establish a northern empire under a succession of dynamic leaders was now reduced to a relatively minor polity.

- Conversely, Poltava also marked the rise of Russia as an important international power that would continue to play a major role in European politics for the next three centuries. The battle solidified Peter's position and enabled him to complete his process of modernizing Russia and incorporating it into the European economic and political sphere.

Suggested Reading

Englund, *The Battle That Shook Europe.*

Konstam, *Poltava 1709.*

Massie, *Peter the Great.*

Questions to Consider

1. In what ways did the introduction of gunpowder to the battlefield, as exemplified by the Battle of Poltava, change how wars were fought?

2. Do you think Charles or Peter was the better leader, and why?

1759 Quebec—Battle for North America

Lecture 23

In 1754, an ambitious but inexperienced 23-year-old lieutenant colonel in the Virginia militia was dispatched with a small contingent of soldiers to a disputed area of western Pennsylvania. There, he was to locate French troops who were constructing a fort and order them to depart from what the British considered to be their territory. When his demands were ignored, the officer attempted a sneak attack that failed miserably, and he was forced to withdraw. This incident would be insignificant but for two facts. First, the young lieutenant colonel's name was George Washington; second, the episode initiated a chain of events that helped spark a war between England and France.

Backdrop to the Battle

- By the mid-1700s, England's American colonies were firmly entrenched along the Atlantic seaboard with a population of approximately 1.5 million. But colonial expansion was hampered by the French conviction that the land west of the Appalachians was theirs.

- Although the French in North America numbered just 70,000, they laid claim to a vast diagonal band of territory stretching from the St. Lawrence River through the Great Lakes to New Orleans. Daring *voyageurs* traveled westward through the Great Lakes and explored south along the Mississippi and Ohio rivers, establishing a series of trading posts and forts along the way and blocking westward movement by the English.

- After a series of escalating border incidents, in 1756, war broke out between France and England. Among the British objectives was to seize control of the strategic St. Lawrence waterway. Situated on a series of high bluffs overlooking this river was the capital of New France, Quebec. If Quebec could be taken, France's main route of communication and trade would be severed, and the British could lay claim to North America.

The Opponents

- James Wolfe was from a military family, and from a young age, he dreamed of achieving distinction in the army. By virtue of both his abilities and his aggressive self-promotion, Wolfe ascended through the officers' ranks of the British army with unusual rapidity—by the time he gained command of the attack on Quebec as a major general, he was only 32.

© chrisdorney/iStock/Thinkstock.

James Wolfe continues to be a polarizing figure; he is portrayed in biographies as being everything from one of the great military geniuses in history to a borderline incompetent.

- Wolfe's opponent at Quebec was the Marquis de Montcalm, a French nobleman also descended from a long line of military officers. Like Wolfe, he was an experienced soldier who was demonstrably brave on the battlefield.

- Although possessing a charm and sophistication that Wolfe lacked, Montcalm had some of the era's prejudice against those he regarded as social inferiors. Thus, when assigned to command in North America, he openly expressed his disdain for both the Indians, who were allies of the French, and the local French Canadians.

- The Canadian-born governor-general of New France had an especially thorny relationship with Montcalm and disagreed radically with him over how the war should be fought. The governor-general advocated a kind of guerilla warfare using Indians and local soldiers that Montcalm found morally troubling.

The Campaign

- Quebec is situated on a high, rocky promontory on the north side of the St. Lawrence waterway. It is protected to the east and south by the St. Lawrence and to the north by the St. Charles River. The walled Upper City stands high above the river on a plateau; the smaller Lower City is below, by the river. On the plateau west of the city spread the flat Plains of Abraham. Along the edges of this plateau are steep cliffs dropping 200 feet to the St. Lawrence.

- The French believed that their cannons could prevent British ships from proceeding up the St. Lawrence River. To prevent British troops being landed further down the St. Lawrence and marching up to Quebec, Montcalm had an elaborate series of fortifications and strong points constructed along the northern escarpment, ending near a 300-foot cliff over which the Montmorency Falls plummeted. Montcalm nevertheless judged that this zone of the St. Lawrence was the most likely area for a British landing; thus, he stationed the majority of his field army here.

- To capture this stronghold, Wolfe had about 8,500 men. What it might have lacked in numbers, his force made up in quality: 10 excellent battalions of British regular line infantry supplemented by some grenadiers, a contingent of artillerymen from the Royal Regiment of Artillery, and 6 companies of American rangers.

- Wolfe's main subordinates were all young, eager, and competent. He also enjoyed the support of the sizable fleet of Royal Navy warships and crews that transported his army to its target. This fleet offered an important resource in terms of spare cannons, gunpowder, supplies, manpower, boats, and experience with amphibious landings.

- On the other side, Montcalm had perhaps 13,000 to15,000 troops of all kinds, but the core of his forces was his 8 battalions of regular French infantry, although these were badly under strength and probably totaled no more than 5,000 men.

- The British arrived in the region of Quebec in June 1759, and Wolfe determined to launch an assault at the far end of the French fortified line on the northern bank, near the Montmorency Falls. The British attacked in the late afternoon of July 31.

- The British grenadiers managed to capture one French redoubt but could not force their way into the main French lines. Another group coming along the shore bogged down, and the assault was called off. In addition to failing to achieve its objectives, the attempt cost Wolfe more than 400 prime soldiers.

- But the British had discovered that they could run ships upriver past the guns of Quebec, and Wolfe had begun a two-pronged strategy designed to wear down the defenders' will to resist: a combination of direct bombardment and a campaign of laying waste to the surrounding countryside.

- Employing heavy cannon and mortars, Wolfe pounded the city without mercy. The shelling sparked destructive fires, and large sections of Quebec were effectively flattened. Out in the countryside, an estimated 1,400 farms were burned, and the region around Quebec was transformed into a smoking wasteland.

The Battle

- Wolfe chose a direct assault aimed at the cliffs west of the city. Montcalm had established a series of pickets to keep watch along these cliffs, but most people believed that they were too steep to offer a reasonable route of attack. Wolfe had noted one spot where a steep trail ran to the top. The trail was blocked by a barrier, there was a strong guard post at the top, and a nearby battery covered the landing site at the base of the trail, but Wolfe decided that this spot constituted a weak point in the defenses. He decided on a nighttime landing from boats that would seize control of this trail.

- The plan was for the assault force to start upriver and drift down silently on the current, land at the base of the trail, slaughter the guards, fight their way to the top, and overwhelm the guard post there. The main army could then follow and strike Quebec from the land side. Several other diversionary attacks would be made the same night.

- The British now enjoyed a streak of luck that allowed the attack force to seize control of the vital trail.
    - Montcalm had stationed a French officer with a powerful flying column of elite troops opposite the British ships from which the attack was to emanate. The officer was to keep a close watch on the ships and move to intercept and destroy any boats launched from them. On the critical night, he failed to notice the heavily laden boats gliding downstream on the current.

    - The officer in command of the detachment guarding the trail, Captain Vergor, had been told to expect a convoy of French boats that night bringing supplies to the city. When the British boats materialized out of the darkness, the French sentries naturally took them for the supply convoy.

    - Some of the English boats, carried on a strongly ebbing tide, overshot the landing zone by several hundred meters and came ashore at the base of the cliffs where there was no apparent route. The light company under the command of Colonel Howe threw themselves at the cliff face and scrambled directly upward. Coming from this totally unexpected direction, they took the sentries guarding the top of the trail totally by surprise.

- Attacked on two sides by Howe's men coming along the cliff and by other British troops storming up the trail, Captain Vergor's men were overwhelmed or fled. Wolfe and the rest of the British army were now able to land and climb up the trail.

- Dawn found Wolfe and the British army of around 3,000 men drawn up in a line of battle facing the city. The French opened fire at relatively long range, and following their usual practice, individual soldiers shot and reloaded as fast as they were able. The British, by contrast, held back until the French were 40 to 60 yards away, then fired by platoons.

- The battalions at the center of the British line held their fire even longer, then unleashed a single, more coordinated volley. Wolfe had ordered his men to load their muskets with an extra ball, and observers recorded that the effect of this one volley was devastating to the French ranks. In just a few minutes, it was all over. The city surrendered five days later, on September 18.

Outcomes

- The fall of Quebec was a decisive moment. The battle directly contributed to the birth of the United States because the British crown levied new, heavier taxes on the American colonies to pay for it: a source of considerable resentment and one of the principal causes of the American Revolution.

- On the French side, the war weakened the monarchy and drained the treasury, provoking harsh new taxes that contributed directly to the French Revolution.

- In the long term, the removal of French power along the Mississippi River and other regions in the Midwest allowed the newly independent American colonies to expand beyond the Appalachians and claim the rest of what would become the continental United States.

Suggested Reading

Hibbert, *Wolfe at Quebec.*

Reid, *Quebec 1759: The Battle That Won Canada.*

Stacey, *Quebec 1759: The Siege and the Battle.*

Questions to Consider

1. In how many ways did luck play a role in the outcome of the battle discussed in this lecture?

2. What do you think the long-term effects would have been if Wolfe's army had been repulsed from its attack on Quebec?

1776 Trenton—The Revolution's Darkest Hour

Lecture 24

In December of 1776, the rebellion of the American colonies against their British overlords looked to be coming to a dismal and premature end. The colonists had scored some promising early victories, but by the fall of 1776, the British Empire had struck back with a vengeance. Washington was forced to retreat through New York and New Jersey, and the British captured several key rebel fortresses. By that winter, many believed that the Revolution was doomed to failure. Yet from this low point—in just 10 days between December and January—the colonists would gain an irresistible momentum, thanks to the bold maneuvers of Washington and the battles at Trenton and Princeton.

Dark Days of the Revolution

- In the opening months of the American Revolution, everything had seemed to go the way of the colonists. Indeed, when the British army was forced to flee from Boston in March of 1776, many thought that the war was effectively over. This triumph was followed during the summer by the emotional high point of the signing of the Declaration of Independence.

- In August of 1776, the British commander-in-chief, Howe, launched an assault on New York and defeated the Americans at Brooklyn Heights, the largest battle yet. George Washington, the leader of the rebel army, was forced to flee with his forces to New Jersey.

- In October, the British and Americans fought another major battle, but once more, the colonists were defeated and had to fall back. This process was repeated throughout the fall, as the British chased Washington south through New Jersey, winning a string of victories and swiftly capturing several key fortresses that the rebels had depended on to retard the British advance.

- By December 1776, when both sides settled into winter camps, Washington had lost 90 percent of his army to death, desertion, division, capture, or simply because the volunteer soldiers' short terms of enlistment had expired.

- Washington's reputation as a good general had been erased by that fall's unrelieved string of demoralizing defeats and retreats, meaning that no new recruits would show up for the next campaign season in the spring. To observers on both sides that winter, it seemed certain that the American experiment was coming to an unsurprising end as a miserable failure.

- Yet between December 25, 1776, and January 3, 1777, the elite British soldiers would be defeated not once but twice; Washington's reputation as a military genius would be reestablished; and from this low point, the revolution would gain an irresistible momentum, culminating in the defeat and surrender of the British at Yorktown and the acknowledgment of independence for the fledgling American Republic.

The Opponents

- The person responsible for this turnaround was Washington himself. As a general, he had displayed a talent for holding together the often unruly and independent volunteers who made up the American ranks, and on the march, he habitually hovered at the end of the line to give encouragement to those who were lagging.

- The general commanding the British army was an aristocrat named William Howe. At the Battle of Quebec, he was the daring officer who ordered his men to scramble directly up the steep cliffs, taking the French by surprise and winning the battle. Howe was a political moderate who disagreed with the Crown's harsh treatment of the American colonies and who had considerable sympathy for them. Still, he firmly believed in the British Empire and its divinely mandated mission to rule.

- Howe's older brother, Richard, had a similar career arc, but in his case, it was in the navy. Sent to sea at 13 as a midshipman, he, too, had risen quickly through the ranks, gaining a reputation for boldness and personal courage. He evinced an unusual degree of concern for the well-being of the sailors under his command and won a series of notable victories, becoming the youngest admiral in the fleet.

The password chosen by Washington for the crossing of the Delaware plainly reflected what he believed was at stake: "victory or death."

- When the conflict with the American colonies broke out, King George III personally asked the Howe brothers to take charge of Britain's land and sea forces in North America and to resolve the dispute by peaceful means if possible but by force if necessary.

- Another important figure was General Charles Cornwallis. Much like Howe, Cornwallis was of the highest aristocratic stock, and he, too, served with distinction in the Seven Years' War. It was his army that chased Washington across the length of New Jersey.

- The soldiers employed by Britain against the Americans were an experienced group of regular line British infantry regiments supported by grenadier companies, light companies, mounted dragoons, and mobile artillery units. The infantry typically fought deployed into a long line two ranks deep.

- Howe's army also included a large contingent of Hessian regiments. These were mercenaries recruited from the Hesse-Kassel region of Germany. Ultimately, around 20,000 Hessians, plus another 10,000 men from other regions of Germany, would enlist to serve in the war against the American colonies. They were highly disciplined and formidable troops.

The Battles

- When Washington reached the Delaware River, he ferried his men across to Pennsylvania, then had them gather all the boats and watercraft they could find and place them on the south shore, thus denying the British a means to cross the river and continue their pursuit. This stratagem, coupled with the onset of winter, temporarily halted the British, who now settled down in New Jersey, waiting for spring.

- In these extreme circumstances, Washington decided on one last desperate attempt before his men's enlistment expired with the turn of the new year. As a target, he selected the contingent of about 1,500 Hessians settled for the winter in Trenton, who were under the command of an experienced soldier, Colonel Rall.

- Washington planned a three-column assault in which his army would cross the Delaware at nightfall on Christmas Day and converge on Trenton before dawn. Washington would personally lead one of the attacking columns. Although the weather on Christmas Day had been clear, as dusk fell, a northeaster struck, bringing rain, snow, and biting winds.

- The difficulty and complexity of moving such a quantity of men, animals, and equipment at night and in such appalling conditions inevitably caused disruption to Washington's plans. The crossing took far longer than he calculated, and the sun was already close to rising by the time Washington's force was assembled and ready to begin the hike to Trenton. Worse, the weather prevented the other two columns from crossing at all. Washington determined to press on and divided his force into two sections for the march.

- When the columns reached the outskirts of Trenton, they overran the outlying Hessian pickets. Six American artillery pieces were deployed to fire down the length of the two main streets, and the attack began.

- The well-sited American cannons played a key role in dispersing Hessian attacks and in keeping them from organizing. As the Hessians began to be overwhelmed and fall back, Rall was mortally wounded by two musket balls. The final casualty figures were about 100 Hessians killed and wounded and more than 900 captured.

- Though militarily of marginal significance, Washington's daring and successful raid immediately captured the American public's imagination and rekindled enthusiasm for the war. Howe immediately dispatched Cornwallis, who had not yet departed on his ship for England, back to New Jersey to take charge.

- Washington was determined to compound the momentum by launching an even more ambitious offensive, but he required additional manpower. He was able to consolidate under himself troops from other commanders, but he also needed to convince enough of his own veterans not to depart as their enlistment expired. He was able to retain about 1,400 of his experienced men, giving him a solid core of veterans for the coming campaign. They crossed over the Delaware to Trenton again and took up defensive positions along a ridge near Assunpink Creek.

- Intent on avenging the embarrassing defeat of the Hessians at Trenton, Cornwallis stormed through New Jersey, gathering forces as he went. As he approached Trenton, he had an army of more than 9,000 men, and he ordered one brigade of about 1,000 under Lieutenant Colonel Mawhood to hold position as a reserve at Princeton. On January 2, he engaged the American line at Assunpink Creek. Fighting from the woods and prepared positions, the Americans stubbornly resisted the British advance, throwing back several attacks, but steady pressure forced them to give ground.

- As night fell, Cornwallis called off the assault, confident that the Americans were close to breaking and that when the battle resumed in the morning, he could turn Washington's right flank and secure the victory. Some of Cornwallis's aides urged him to finish off the Americans immediately, but believing that he had Washington trapped, Cornwallis rejected the advice.

- Washington was indeed deeply worried that the next day's battle would see his army caught with their backs to the river. Again, he decided on a bold nocturnal maneuver that would surprise and outwit his enemy. He gave orders for his army to sneak away from Trenton during the night, to march in a dangerously exposed path all the way around Cornwallis's sleeping army, and to fall upon the British reserve left at Princeton.

- While one group stayed behind to keep the campfires lit, the bulk of Washington's army slipped away. The cannon wheels were muffled with cloths, and the soldiers were ordered to observe strict silence. The army remained undetected, and by dawn on January 3, the leading elements were approaching Princeton.

- The Battle of Princeton was much more of a conventional European-style stand-up fight between two opposing lines of men than either Trenton or Assunpink Creek had been. Across an open field, the two sides exchanged a series of volleys at close range, followed by charges and countercharges.

- At one point, Washington, conspicuous on his grey horse, rallied the faltering American lines by dangerously exposing himself to British fire at a range of less than 50 yards. Eventually, the Americans prevailed against the outnumbered British, and Washington scored yet another brilliant victory. In some ways, this one was even more inspiring because it had shown that the American militia could go toe-to-toe with the British regulars and defeat them.

Outcomes

- The Revolutionary War continued for several more years, until it finally ended with Washington's victory over Cornwallis at Yorktown in 1781. But the true turning point was the twin victories of Trenton and Princeton. These came at the very darkest moment for the Revolution, at a time when support for it was faltering and everything seemed to be turning against the young republic.

- Instead, by a series of clever and daring maneuvers, Washington restored the enthusiasm and confidence of his army, firmly established his reputation as a general, and saved the nascent revolutionary movement. From that moment on, the continuance of the Revolution was never again seriously in doubt, leading inexorably toward the colonists' attainment of freedom from Great Britain.

Suggested Reading

Bonk, *Trenton and Princeton 1776–77*.

Dwyer, *The Day Is Ours!*

Fischer, *Washington's Crossing*.

Questions to Consider

1. What were Washington's positive qualities as a commander, and which of them do you think was most important?

2. If Washington had lost the Battle of Trenton, do you think that the United States would eventually have achieved independence from Britain?

1805 Trafalgar—Nelson Thwarts Napoleon

Lecture 25

On January 8, 1806, an unprecedented funeral was held in England. The casket was conveyed down the Thames in a royal funeral barge; the next day, it was transferred to a hearse shaped like a warship that sailed through the streets to St. Paul's Cathedral. Tens of thousands of weeping onlookers filled the cathedral and lined the roads. The man who had been granted these honors was a warrior who had won a spectacular and crushing victory over a superior force and delivered his nation from years of desperate fear of invasion by an implacable and terrifying foe. At the very moment of his triumph, this hero, Horatio Nelson, had been struck down in battle.

Naval War in the Age of Sail

- Since the galleys had clashed at Lepanto, naval warfare had undergone rapid change.
  - Deep-hulled ships propelled by square-rigged sails that could tack against the direction of the wind were capable of carrying enough supplies to cross the oceans.
  - When heavy cannons were mounted on these ships, the new technology marked a new era of naval warfare.
  - But only one or two cannons could be mounted at the front or stern of a ship; thus, guns were positioned mainly along the sides. Ships could not shoot in the same direction that they were moving and could not fire more than half their guns at any one enemy ship.
- In the line-of-battle formation, the fleet of one adversary would line up in a long column, all sailing in the same direction. The two opposing lines of warships sailed in opposite directions, firing broadsides at one another. The two sides simply pounded one

© AlamarPhotography/iStock/Thinkstock.

The *Victory*, Nelson's flagship at Trafalgar, was one of the largest English warships—200 feet long and 50 feet wide and carrying 102 guns.

another until a ship sank, caught fire, or suffered so many casualties that it had to surrender.

- A typical warship of this period had two or three decks for guns. The largest warships carried between 64 and 120 guns. These warships represented an enormous investment of money and resources to construct, were extremely costly to operate, and incorporated the most cutting-edge technology of the day.

England versus France

- The dominant power in naval warfare in the 18th and 19th centuries was England. This nation possessed a large number of excellent natural harbors from which the main sea routes to Europe could be controlled. England developed the largest merchant marine in the world, as well as the largest and most efficient navy to protect it.

- England's greatest enemy during this time was France, which had just undergone the chaos of the French Revolution and had fallen under the rule of an ambitious military genius, Napoleon Bonaparte. Through a series of astonishing and swift campaigns, Napoleon extended his dominion over much of Europe, from Spain to the borders of Russia. The instrument of his success was his Grand Army, a highly trained, unified force that, at its peak, numbered more than 700,000 men.

- By 1804, Napoleon's army stood poised along the French coast, and he had thousands of barges constructed to ferry his troops across the English Channel. What stood in his way were the warships of the English navy.

- Against this menace, the English strategy was to preserve control of the seas by not allowing the French navy to leave its harbors. Outside of every major French harbor, a squadron of English ships kept up a constant blockade, maintained even during storms, placing an enormous strain on the ships and their crews.

The Opponents

- Many of the ships and men that sustained this vigil against the French and fought in the war at sea have become legendary, but none more so than Horatio Nelson. He was born in 1758 to a prosperous family and was sent to sea at the age of 12 as a midshipman.

- Nelson rose through the ranks to lieutenant and eventually captain and fought in a number of battles, in which he established a reputation as a particularly innovative and aggressive officer. He was also one who led from the front, a practice that took a physical toll: In one battle, he lost an eye, and in another, his right arm was shattered by a musket ball and had to be amputated. The loss did not impair the success of his career, however.

- By 1798, Napoleon had conquered most of Europe and decided to lead an expedition to Egypt. He amassed a large fleet of warships and transports to carry his army and personally took charge, sailing to Egypt, landing near Alexandria, and subduing the country after fighting several battles. While Napoleon fought in Egypt with the army, his fleet anchored in a line close to the shallow water in a strong defensive position near the mouth of the Nile at Aboukir Bay.

- Nelson was sent with a smaller force of ships to attempt to destroy the French fleet and strand Napoleon in Egypt. In a typically aggressive fashion, he sailed to Aboukir Bay, arriving near dusk and, rather than waiting for the next day, attacking immediately.

- The action that followed is known as the Battle of the Nile. The French vessels were larger and more heavily armed, but the British were better trained and attacked enthusiastically. The most dramatic moment of the battle came when the French flagship (the *Orient*), an enormous three-decker with more than 100 guns, caught fire, and the gunpowder magazine exploded, blowing the ship apart. By the morning, the English fleet had won a decisive victory, and the majority of the French ships were either captured or destroyed.

- This was an important victory for England. Without the support of his fleet, Napoleon was forced to give up the expedition. In the end, he had to abandon most of his army in Egypt and sneak back to France alone. Although Napoleon might be dominant on land, as long as England controlled the sea, his dreams of world conquest would remain unrealized.

The Battle

- By the middle of 1805, the English blockade had been going on for years, and Napoleon was getting tired of waiting. By capturing Spain, he had also gained control of the Spanish fleet, and his new plan was to combine the French and Spanish fleets into a force so large that it could overwhelm any English opposition.

- This combined fleet would sweep up from Spain, push aside or destroy the English ships in the channel, and give his army the chance it needed to cross. The admiral put in charge of this mission, Villeneuve, amassed a combined fleet of 33 ships of the line, including a number of particularly large battleships, including the biggest, most powerful ship in the world, the colossal four-deck, 136-gun *Santissima Trinidad*. On October 20, 1805, the combined fleet set out from Cadiz.

- Nelson was charged with stopping the fleet, and he had 27 English ships to oppose it, including his flagship, the 102-gun *Victory*. At a meeting of his captains, Nelson outlined a bold new strategy: Rather than using the line of battle, Nelson planned to form his ships into two shorter lines perpendicular to the French and charge straight at them, cutting the French line of battle in two places. His ships would then cluster around those of the enemy and attempt to overwhelm them.

- The two fleets met off Cape Trafalgar in Spain. It was the biggest battle of the age, involving 60 ships of the line and 50,000 men. As Nelson sailed toward the enemy, he used signal flags to send a message to his crews that has since become one of the most famous in military history: "England expects that every man will do his duty."

- The battle unfolded much as Nelson had intended, with the two columns of English ships having to endure the combined fleets' fire as they approached. Nelson and the *Victory* led one of the English lines, and one of his trusted captains led the other. Both succeeded in breaking the combined fleets' line of battle, and the conflict turned into a confused melee of ships blasting at each other at point blank range.

- As the battle raged, the *Victory* was surrounded by several enemy ships. Nelson was pacing back and forth on the deck when a musket ball shot from the mast of the French ship *Redoubtable* struck him in the shoulder, passed through his chest, and lodged in his spine. Nelson immediately realized the wound was fatal, and he died later that day. He lived long enough, however, to hear the news that his fleet had won a crushing victory over the enemy.

- Of the French and Spanish ships, all but 11 were captured or sunk. The combined fleet suffered 7,000 casualties. On the English side, not a single vessel was lost, though many were heavily damaged, and there were 1,600 casualties.

Outcomes

- Trafalgar was one of the keys to defeating Napoleon's goal of universal domination. Although Napoleon's power was as yet unchallenged on the continent, Trafalgar put an end to his plans to invade England. By ensuring England's survival, the battle guaranteed that there would always be a strong European nation that could and would provide a focal point for opposition to Napoleon.

- After Trafalgar, Napoleon pulled his armies away from the English Channel and instead committed them, disastrously, to the invasion of Russia. Although it would still take much to defeat Napoleon, Trafalgar was the battle that checked his ascendency and marked the beginning of his decline.

Suggested Reading

Adkins, *Trafalgar: The Biography of a Battle.*

Cannadine, *Trafalgar in History.*

Knight, *The Pursuit of Victory.*

Pocock, ed., *Trafalgar: An Eyewitness History.*

Questions to Consider

1. Which of the many technological advancements mentioned in this lecture do you think most revolutionized naval warfare and why?

2. How much credit for the victory at Trafalgar do you think was due to Nelson and how much to the general superiority of the English sailors in terms of training and experience?

1813 Leipzig—The Grand Coalition

Lecture 26

What was the largest battle fought in Europe before the First World War? The answer is the Battle of Leipzig, which took place in 1813 and involved more than half a million soldiers. It was a gigantic clash that included the armies from most of the leading nations of Europe at the time. More than any other battle of the era—even the much more famous Battle of Waterloo—Leipzig was the decisive moment when Napoleon's dreams of European domination were finally defeated.

The Campaign

- After the defeat of the combined French and Spanish fleets at Trafalgar destroyed Napoleon's hopes of invading England, he turned his attention to new conquests on the mainland, cleverly exploiting longstanding rivalries and resentments among his potential enemies to keep them from uniting against him.

- Eventually, Napoleon focused on Russia as a target, and in 1812, he led his army of more than half a million men into Russia. Although Napoleon won several battles and even managed to capture Moscow, his frozen and debilitated army was eventually forced to retreat, suffering severe hardships and more casualties. The campaign ended with the almost total annihilation of Napoleon's army.

- Back in France, Napoleon rebuilt his army with astonishing speed. The battle-hardened survivors of the Russia campaign gave him a solid core of officers and veterans. To amass sufficient numbers of new soldiers, he had to draw on every possible source, even calling up several classes of young men ahead of schedule. By these means, he was apparently able to bounce back from the brink of disaster in a remarkably short time.

- On the surface, it looked as if Napoleon was now just as strong as ever, but there were several key underlying weaknesses in his new army.
    - First, the new conscripts truly represented the bottom of the manpower barrel—if Napoleon lost these men, there would be no replacing them.

    - Second, he had lost nearly 200,000 horses and most of his cavalrymen in Russia, and it proved to be much more difficult to replace these trained horses and riders than to find new infantry recruits. Thus, when Napoleon's new army went to war, it would be badly under strength in cavalry.

- One of the most important roles of cavalry in Napoleonic warfare was to run down and wipe out a defeated enemy to ensure that it did not survive to fight another day, but at key moments over the next several years, Napoleon would be unable to finish off a beaten foe for lack of adequate cavalry.

- Another consequence of Napoleon's defeat in Russia was the loss of several powerful continental nations that, up to this time, had been his reluctant allies; these nations now felt safe to desert Napoleon and join up with the enemies of France. The most significant of these was Prussia.

- The Prussians signed an agreement with England, Russia, Spain, Portugal, and Sweden to work together in a joint effort to bring down Napoleon. This alliance was known as the Sixth Coalition—its very name a testimony to how many times Napoleon had survived international attempts to unseat him.

- The great question remaining was which side Austria would join. The opposing alliances were so closely matched that whichever one could add Austria's large army to its strength would have the advantage. Throughout the late spring and summer of 1813, the Austrians waffled while the fate of Europe hung in the balance.

- Russia and Prussia moved aggressively, sending a combined army against Napoleon's numerically superior force. In two battles, Napoleon defeated them, each time compelling the combined army to retreat, but partially because of his lack of cavalry, he was not able to destroy them or take the initiative himself. Thus, although these battles technically counted as victories, they also revealed Napoleon's weaknesses. This outcome sufficiently emboldened Austria to side with the Sixth Coalition.

- In the late summer and early autumn of 1813, Napoleon sparred with coalition forces all across greater Germany. He won a clever victory at Dresden, but this was balanced out by the defeat of nearly all his marshals in other battles. Again, Napoleon's ability to track the various coalition armies was severely hampered by his shortage of cavalry. Eventually, the remaining French forces, numbering about 185,000 soldiers, coalesced around Napoleon at Leipzig.

- Closing in on the city were four separate coalition armies, totaling approximately 330,000 men. The largest of these was a Russian-Austrian force of about 150,000: the Army of Bohemia. The others were the predominantly Russian Army of Poland with about 50,000 men, the Army of the North with 65,000, and the Army of Silesia with another 65,000.

The Battle

- The Battle of Leipzig lasted from October 16 to October 19, unfolding in a number of separate stages. On the morning of October 16, Napoleon launched his main offensive, which he personally oversaw, against the Army of Bohemia to the south. Meanwhile, he ordered Marshal Marmont to conduct a holding action against the allied forces coming toward Leipzig from the north.

- Much of the fighting centered on the village of Wachau and its surrounding fields and woods. The ferocious duel in this sector seesawed over the course of the day, with ownership of the village changing hands no fewer than three times.

- In the early afternoon, after a particularly intense French barrage had weakened the Russian lines, Napoleon ordered Murat to charge at the head of 10,000 cavalry. The Russians managed to bring up reserves to blunt the attack, but as night began to fall, Napoleon's forces had pushed deep into the enemy lines and gained a considerable amount of territory.

Napoleon probably should have retreated on October 17, but he was intent on securing the victory that he believed would lead to the breakup of the Sixth Coalition; thus, he stayed and drew in his forces tighter around Leipzig.

- Although the Army of Bohemia was battered and driven back by Napoleon's day-long assault, in the end, its lines did not break. Napoleon's great offensive had failed and, along with it, his best chance to win the Battle of Leipzig.

- Meanwhile, Marmont was desperately trying to stave off the attack of the Prussian General Blucher and his Army of Silesia in the north. Again, several of the villages outside Leipzig became focal points of the struggle, especially the hamlet of Mockern, which served as a French strongpoint. Finally, Marmont was wounded in a Prussian cavalry charge, and Blucher captured Mockern.

- On October 18, the coalition launched an all-out assault, with all the armies advancing on the encircled French. Their plan was simple: Attack until victory was achieved. In the afternoon, the tide began to turn against the French. As night fell, it was clear that the French had lost the battle. Napoleon himself was described as being in a depressed state, and inadequate preparations had been made for a retreat from Leipzig. The only remaining question was whether the French could escape the ring that was closing in around them.

- Napoleon had suffered a significant defeat, and his army had taken a severe battering. His forces were defeated but not broken, and the regiments were still disciplined and in good order. If he could retreat from Leipzig with his remaining men and equipment intact, the French might yet fight another day with a large enough army to challenge the Sixth Coalition and perhaps pull out another brilliant victory.

- Because the French were surrounded on three sides by the coalition forces, their only escape route was to the west, over causeways that traversed some marshy territory. To reach this road, the fleeing army first had to cross a bridge over the Elster River. All morning, French forces streamed over this bridge to safety.

- At this moment, however, an incident transformed the battle from a discouraging but potentially survivable defeat into a fatal disaster. Napoleon had put one of his generals in charge of the Elster bridge with orders to blow it up after the French army had crossed. This general delegated responsibility to one of his colonels, who placed the explosives. Because this colonel was unsure which unit was the rearguard, he went to headquarters to find out, leaving a corporal behind.

- When a few Russian skirmishers began shooting at the troops streaming across the bridge, the corporal panicked and lit the fuse. The explosion killed hundreds of French troops and trapped tens of thousands of French soldiers on the wrong side of the river, along with most of the wounded and a sizable portion of the French artillery. Ultimately, more than 40,000 prime troops were stranded and taken prisoner, along with 300 irreplaceable cannons.

Outcomes

- These losses doomed Napoleon. The allies had won a truly decisive victory and one from which the French could not recover. The shaky coalition now had a shared victory and could see a clear path to the end of the war. The coalition forces entered Paris on March 30, 1814.

- Napoleon abdicated and went into exile on the island of Elba, but in February 1815, he escaped from the island, landed in France, proclaimed himself emperor, and began to rally his veterans, initiating the period known as the Hundred Days. Immediately, a Seventh Coalition was formed, and each of the major powers promised to supply 150,000 men to oppose Napoleon.

- The first to get going were the English under Lord Wellington and the Prussians under Blucher. Realizing that his only hope was to defeat his enemies before they could unite, Napoleon moved against them. The result was the Battle of Waterloo, at which Wellington and an army of about 80,000 narrowly defeated Napoleon and his 70,000.

- Waterloo is often cited as the decisive battle of the Napoleonic Wars, and it was certainly the last major battle in that extended conflict. But it was Leipzig that taught Europe how to defeat Napoleon—the quick response of the Seventh Coalition demonstrated that Europe had learned that lesson well—thus, the outcome of the Hundred Days was never really in doubt. Defeated again, Napoleon was shipped off to the island of St. Helena and died there in 1821.

Suggested Reading

Brett-James, *Europe against Napoleon.*

Hofschroer, *Leipzig 1813.*

Maude, *The Leipzig Campaign, 1813.*

Questions to Consider

1. Do you agree with the argument that Leipzig was more decisive than Waterloo? Why or why not?

2. Had the Elster bridge not been blown up prematurely, could this battle still be considered decisive?

1824 Ayacucho—South American Independence
Lecture 27

In June of 1806, a British admiral named Popham, having fulfilled his orders to capture Cape Town in Dutch South Africa, decided to cross over to South America and capture the Spanish viceroyalty of the Rio de la Plata. At first, the locals were glad to see the Spanish evicted, but resistance movements soon appeared. After a year of campaigns, the British were expelled, and Spanish rule was restored. From this conflict, the locals learned two important facts: Spain was too weak to effectively defend them, and they could defeat European troops. When coupled with ideas of freedom and self-rule derived from the recent American, French, and Haitian revolutions, these insights would lead to the emergence of independence movements in South and Central America.

The Liberators

- Among the many leaders who played roles in the Latin American independence movements, Simon Bolivar is prominent for his involvement in the independence struggles of several regions. He grew up in Caracas, where his family represented the extreme upper class. By 1800, his family had already been in the New World for more than 250 years; they were aristocrats whose great wealth derived from, among other things, mines operated using slave labor.

- Bolivar attended a military academy, then went to Europe, where he spent several years in Paris. In Europe, he at first appears to have led the rather typically debauched life of a young, rich, aristocratic playboy, but he was also exposed to the ideas of the Enlightenment and experienced a political awakening. He began to apply the ideals of freedom and self-government that he was hearing and reading about to his home continent.

- When he returned to Venezuela in 1807, Bolivar brought a conviction that his homeland must achieve independence from Spain. Initially, not many agreed, and for some years, he suppressed his radical inclinations. Revolution was inevitable, however, and he became involved in an early attempt to establish a Venezuelan state that became known as the First Republic. Bolivar's performance was not impressive—he suffered a bad military defeat and betrayed one of the leaders of the movement to the Spanish.

- His next attempt was a new uprising accompanied by several written manifestos that reflected Enlightenment ideals. But once again, success was short-lived, and after a counterrevolution, Bolivar departed for a period of exile in Haiti and Jamaica. Upon his return in 1816, the revolutionary movement began to gather greater momentum.

- An early turning point was the Battle of Boyaca in 1819, at which Bolivar's forces defeated a royalist army, leading to the independence of the region occupied by modern Colombia. This was followed in 1821 by another important victory at the Battle of Carabobo, which enabled the creation of a Gran Colombia encompassing the northern quarter of South America. Bolivar himself became president of this new state.

- One of Bolivar's subordinates was a promising officer named Antonio Jose de Sucre; in 1822, he led an army into Ecuador. After a dramatic battle on the slopes of a volcano, Sucre was victorious and soon after captured Quito. As a result of Sucre's actions, Ecuador now joined the list of independent South American countries.

- Meanwhile, revolutionary movements were breaking out across South America. The most important figure in the southern rebellions was Jose de San Martin. Although he was born in the New World, his family moved to Spain when San Martin was just a child. His father was a soldier, and San Martin followed him into the army, serving in the fighting in Spain during the Napoleonic Wars.

- San Martin resigned his commission and returned to South America, where he became involved in the war for Argentine independence. He began to rise to prominence after he organized a unit called the Army of the Andes and, with it, crossed the supposedly impassable mountains and invaded Chile. After a series of battles, Chilean independence was achieved in 1818, winning San Martin considerable renown. Having helped Argentina and Chile gain freedom, San Martin now focused his efforts on attempting to extend the liberation movement to Peru.

British forces, specifically the Albion Legion, played a key role in the battle for South American independence; Bolivar himself once attributed at least part of his success to the recruiting agent in London.

- The two great liberators of the north and the south met at Guayaquil in July 1822 to plan a joint assault against Peru, but they clashed in terms of personality, preferred methods of operation, and, most significantly, long-term goals. Realizing that cooperation between them would be nearly impossible and fearing that even to attempt it might destroy what they had accomplished, the aging San Martin decided to remove himself from the scene. He retired permanently to Europe, where he died in 1850, leaving Bolivar effectively in charge of the South American revolutionary movement.

The Battle

- By 1824, with Columbia, Venezuela, Argentina, Ecuador, and Chile liberated, the primary bastion of colonial Spanish rule in South America was Peru. The conquest of Peru would not be easy because it still contained a strong army of 12,000 royalist troops loyal to the Spanish king, and the highland area of the country was a natural fortress surrounded by the Andes.

- Bolivar and his chief lieutenant, Sucre, began to plan their campaign against Peru. The time was ripe for an attack: Internal turmoil in Spain ensured that the royalist forces in South America would not be receiving reinforcements.

- In May 1824, the revolutionary army embarked on an epic march through the mountains. The soldiers suffered severely from altitude sickness and from nighttime temperatures well below freezing. By August, Bolivar had made his way onto the high Peruvian plateau with an army of nearly 9,000 men.

- On August 6, there was a sharp clash between Bolivar's cavalry and that of the royalists at Lake Junin, at an altitude of 12,000 feet. The royalist cavalry initially had success, but Bolivar's horsemen rallied and drove them off. From a military perspective, this battle was unusual because it was fought solely by the mounted forces using only swords and lances. Although a relatively minor victory, the Battle of Junin greatly elevated the morale of the revolutionary army, and it cleared the way for Bolivar to march into Peru proper, setting the stage for the decisive Battle of Ayacucho.

- Bolivar departed to deal with some urgent political issues, leaving Sucre in charge of the army. The commander of the royalist army opposing Sucre was Jose de la Serna. For several months, Sucre and La Serna shadowed each other. Finally, the two forces drew together on December 8 near the plain of Ayacucho and encamped for the night.

- That evening, Sucre cleverly ordered his musicians forward toward the Spanish, together with some skirmishers. Throughout the night, the band's playing, coupled with the skirmishers firing random shots in the direction of the Spanish campfires, kept the royalist army awake, ensuring that they would be ill-rested for the coming battle.

- The plain of Ayacucho itself was a patch of flat land measuring only 1,300 by 800 yards. Sucre's army of about 6,000 men and one cannon deployed along the western edge. He arranged his infantry battalions in a line, with his cavalry behind them as a reserve.

- On the eastern edge, La Serna deployed his approximately 7,000 men and seven cannons into a similar line of infantry, with some cavalry on each side. As a reserve, he had a large group of cavalry and an elite battalion of halberdiers. He planned to attack and pin down Sucre's army with his right and left flanks, then use his strong center to move in and crush the liberators.

- The battle proper began with the royalist left attacking first and having some success. The rest of the Spanish line also advanced, with Sucre's army coming forward to meet them, and the fighting spread across the length of the lines. The heavier artillery of the royalists gave their attack extra momentum, but Sucre countered with some of his reserves. On the embattled right flank, the commander in charge of stemming the royalist attack firmed up his line and fought them off.

- By afternoon, the battle was turning in favor of the liberators, and the remaining organized royalist forces were falling back into a last-ditch defensive stand on the high ground. La Serna was wounded, and soon after, he and his remaining men surrendered to Sucre.

Outcomes

- With the victory at Ayacucho, the last Spanish resistance crumbled. Ayacucho can be considered a decisive battle in global history because it was the event that clearly ended Spanish rule in Latin America, and it secured and ensured the continued existence of the newly independent South American nations.

- Within a year of Ayacucho, all the territories in the New World had thrown off their colonial overseers and created independent nations. Spain, which had once controlled the greater part of two vast continents, had its New World possessions reduced to Cuba and Puerto Rico. The loss of these territories and of the incomes that came with them caused Spain to plummet from international power and importance.

- Desiring to decrease European influence in the hemisphere, the United States was quick to give diplomatic recognition to the newly formed states and issued the Monroe Doctrine, which declared that any attempt by European powers to reestablish colonial rule in the hemisphere would be considered a hostile act.

Suggested Reading

Chasteen, *Americanos*.

Davis, *100 Decisive Battles from Ancient Times to the Present.*

Lynch, *Simón Bolívar*.

Questions to Consider

1. What do you think were the strengths and weaknesses of Bolivar as a commander, and how did these affect the outcome of history?

2. Do you think the independence of the South American countries was inevitable, and how might history have been different if the independence movements had been delayed even by half a century?

1836 San Jacinto—Mexico's Big Loss

Lecture 28

The Battle of San Jacinto lasted only 18 minutes and involved barely more than 2,000 men on both sides. Compared to many famous battles of history, it would hardly qualify as a minor skirmish, yet the long-term effects of this little clash along the banks of the San Jacinto River were significant, involving, among other things, the transfer of a portion of land larger than the mainland of Western Europe. Even more so than usual, the course of the campaign leading up this battle, as well as its outcome, hinged on the personalities of the opposing commanders, Sam Houston and Antonio Lopez de Santa Anna.

The Opponents

- Sam Houston was one of those remarkable figures of the American frontier whose biography seems too improbable to be true. He was born in Virginia, but his family moved to Tennessee. As a boy, he combined an enthusiasm for adventuring in the woods with adoration of the epics of Homer and Virgil. Despite this love of classical literature, he was a rebellious and indifferent student.

- Enrolling in the U.S. army, he served in the war against the Creek Indians under Andrew Jackson, where his accomplishments included gaining fame for his bravery, attaining the rank of lieutenant, becoming Jackson's protégé, and putting down a mutiny by aiming a cannon at his own men.

- After stints as an Indian agent and a lawyer, Houston embarked on a political career. With Jackson's patronage, he was highly successful, first winning election to Congress, then becoming governor of Tennessee, during which time he shot a man in a duel.

- In 1832, at the age of 39, Houston moved to Texas, gained Mexican citizenship, and became enamored of the region. He soon fell in with the faction agitating for Texan independence and assumed a leadership role among them.

- The commander of the Mexican army at San Jacinto was Antonio Lopez de Santa Anna, a fascinating character who dominated nearly five decades of Mexican history. Daring but vain, gifted but easily bored, capable of manic energy but prone to wallowing in self-indulgence, Santa Anna was obviously a complex individual and continues to be controversial today.

- In 1810, at the age of 16, he enrolled as a cadet in the army. Two months after his enlistment, the Mexican War of Independence began, and Santa Anna spent the next decade variously fighting Indians, insurgents, and royalists and rising to the rank of colonel. He was wounded at least once and cited for bravery and elevated to the rank of general.

- Perhaps his moment of greatest popularity came when Spain made a rather feeble attempt to retake Mexico in 1829, and Santa Anna led the Mexican army in defeating the invading force at the Battle of Tampico. Tampico was a significant victory that solidified permanent Mexican independence from Spain, but Santa Anna reveled excessively in honors resulting from his success.

- In 1833, after several more coups, he became president for the first of what would be more than a half-dozen stints in office. Over the next three years, he was in and out of the presidency three times, eventually dissolving the government and declaring himself dictator.

The Campaign

- New Spain encompassed not only modern Mexico, but it also extended as far north as Colorado and as far west as the Pacific coastline, including all of California. When Mexico gained independence from Spain in the early 1800s, it inherited this vast, sparsely populated region.

- A section roughly equivalent to modern Texas and the state of Coahuila in Mexico was administratively organized into a region known as the province of Coahuila and Texas. The Mexican government was eager to increase population in this province and, thus, encouraged immigrants from the United States to settle there, even granting them Mexican citizenship. This policy proved so successful that 20,000 new American immigrants soon poured into the province, heavily outnumbering the Mexicans settled there.

- In October 1832, the Texan leaders met in a convention and drafted a series of demands to be sent to the Mexican government. The main ones were that the law closing the border to American immigration should be repealed and that the status of Texas within Mexico should be upgraded to full statehood.

- Stephen Austin was dispatched to Mexico City to carry the convention's demands to the Mexican government. By now, Santa Anna was trying to consolidate his dictatorship and suppress revolts against his seizure of power. Austin was accused of treason and thrown into prison. Released three years later in 1835, the former pacifist made his way back to Texas, declaring that war was the Texans' only option.

- These feelings were strengthened when it was announced that Santa Anna planned to personally lead a military force into Texas to punish any rebels. Up to this point, Houston had not played much of a role in politics in Texas, but the threat from Mexico prompted him to call for volunteers to fight Santa Anna, and he offered himself as the commander of this force.

- The next phase of the campaign is well-known. Santa Anna marched an army of about 6,000 men into Texas, toward San Antonio. Rather than withdrawing—as they probably should have done—the local commanders, William Travis and Jim Bowie, decided to stay and oppose the Mexicans, basing their forces in a decrepit Spanish mission called the Alamo. There, a group of around 180 men managed to resist Santa Anna's attacks for nearly two weeks before the garrison was overrun and slaughtered.

- Now, the only remaining opposition to Santa Anna in Texas was Sam Houston and his army of volunteers, numbering around 300. Over the next month, Houston and his men repeatedly retreated, abandoning position after position, as Santa Anna pursued them across Texas. They marched more than 200 miles backward during this period, and Houston turned down a number of opportunities to engage Santa Anna in battle.

- But Houston may have been following a deliberate strategy the whole time. The further Houston went, the longer Santa Anna's lines of communication stretched and the more troops Santa Anna had to detach from his army to garrison key points along the way. Also, the longer Houston delayed, the more time he had to organize and drill his volunteers into a fighting force, and the more volunteers showed up to join the Texan army.

The Battle

- When Houston reached Lynchburg, he apparently decided that the time was right to stop and face Santa Anna. He encamped his army in a line of woods along the shores of Buffalo Bayou. This bayou ran into the San Jacinto River, and the two waterways formed a horseshoe shape enclosing a small plain covered in high grass. The Texans settled down in the tree line on one side to await the arrival of the Mexicans. It has been calculated that Houston's army numbered 930 men, although Houston believed he had only around 780.

- Santa Anna deployed his army on the opposite side of the field. Depleted by disease, marching, and the need to leave men in garrisons, the Mexican army was down to around 950 men and one large cannon, either a 9- or 12-pounder. Santa Anna ordered this gun to be brought up and fired at the Texans, and Houston countered with his cannons; thus, the opening stage of the battle took the form of an artillery duel.

- Houston held a council of war that evening, deciding to wait for the Mexicans to attack first. Santa Anna spent a sleepless night organizing his men and making sure that the sentries were alert against a possible surprise nocturnal assault by the Texans. He had sent orders to his brother-in-law, General Cos, to meet him and to bring along 500 elite soldiers to reinforce his army, and he did not want to engage until these troops arrived.

- To Santa Anna's displeasure, however, Cos brought with him about 400 new recruits, rather than the highly trained, experienced men that had been requested. To get them in fighting shape, Santa Anna ordered the exhausted men to get some food, then take a nap. Tired from being up all night himself, Santa Anna lay down beneath a tree, having given orders that a strict watch be maintained on the Texans and that he himself should be awakened at any sign of movement. Unfortunately, these orders either were not carried out, or the designated sentries failed to perform their duties.

- The Texans grew increasingly restless, and by midafternoon, their frustration reached the boiling point. Houston ordered his men to deploy for battle and took his position at their head. The Texans trotted quickly forward through the tall grass in a long line, dragging the cannon on leather straps. First to reach the Mexican encampment was the left side of the Texan line. They opened fire and were soon joined by the cannon and the rest of the line. The Mexicans finally began to fire back, getting off several shots with their cannon.

- The Texas line swept over the low defensive wall erected by the Mexicans and spread through the camp. Caught totally by surprise, most of the Mexican troops panicked and ran. Santa Anna briefly attempted to rally his men, but realizing it was hopeless, he mounted a horse and fled into the swamps. In just 18 minutes, the camp was overrun and the fighting was over.

Outcomes

- Santa Anna was caught and, in exchange for his freedom, signed two treaties in which he promised to withdraw his troops and work for Texan independence. Texas spent nine years as an independent nation, then joined the United States in 1845. Ongoing disagreement over where the Texas border was located then sparked the Mexican-American War, which ended with the treaty of Guadalupe Hidalgo. Mexico agreed to cede to the United States a further chunk of territory that included modern California, Utah, Nevada, and New Mexico.

- Bitter debates in Congress over whether such new states as Texas and those formed out of the territory gained from Mexico should be slave-owning or not became one of the main issues leading to the secession of the southern states and the American Civil War.

Suggested Reading

Fowler, *Santa Anna of Mexico*.

Haley, *Sam Houston*.

Moore, *Eighteen Minutes*.

Questions to Consider

1. Do you agree with Houston's strategy to keep retreating, or should he have fought Santa Ana earlier?

2. From a Mexican perspective, do you think that Santa Ana was ultimately more of a positive or a negative force?

1862 Antietam—The Civil War's Bloodiest Day
Lecture 29

It could be argued that September 4, 1862, represents the high point of the Confederacy and its attempt to secede from the United States. Over the previous months, southern armies had thwarted the Union's efforts to capture Richmond. Then, at the Second Battle of Bull Run, Stonewall Jackson had inflicted a crushing defeat on the Union army defending Washington, D.C. Riding this wave of victories, the morale of the southern soldiers was at an all-time high, and the Confederate leadership was confident that the South was on the verge of achieving independence from the rest of the Union. Accordingly, on September 4, Lee's Army of Northern Virginia crossed into Maryland for what they hoped would be the final campaign.

Backdrop to Antietam

- After the Second Battle of Bull Run, the route to Washington for Confederate forces now lay open, but Robert E. Lee knew that he did not have the strength to directly assault the defenses of the northern capital. He devised a plan to bypass Washington and instead invade Maryland. This would be the first time that the Confederacy had invaded the north, and it represented a shift from defense to offense.

- Lincoln faced midterm elections in November, and a successful Confederate campaign in Maryland might discredit him and embolden antiwar northerners into pressuring him to seek peace. Furthermore, Maryland was believed to harbor many southern sympathizers, and Lee hoped that the presence of his army there might provoke the state into switching its allegiance, perhaps even triggering other border states to abandon the Union.

- At the same time, major European powers were actively considering granting political recognition to the Confederacy, which would establish the legitimacy of the Confederacy and likely result in vital material aid. Thus, the march into Maryland on September 4 by

Lee's Army of Northern Virginia might be the final campaign of the Civil War.

The Opponents

- Lee embodied the stereotype of the southern gentleman. He came from one of the most aristocratic families in Virginia. His father had been a renowned commander during the Revolution, was a friend of George Washington, and had served as governor of Virginia. At the outbreak of the Civil War, Lee's strong allegiance to his home state caused him to decline a high position in the Union army and instead offer his services to the Confederacy.

- After General Joseph Johnston was wounded on June 1, 1862, Lee was appointed commander of the Army of Northern Virginia, operating in the corridor between Washington and Richmond. One of Lee's best qualities was the confidence and loyalty he inspired in his troops, which enabled him to demand much of them even under conditions of severe hardship. He had a talent for using bold troop movements to outmaneuver and often defeat numerically superior opponents.

- Lee's undoubted abilities as a general were sometimes undermined by a failure to consider the overall strategic picture; as the war progressed, he clung far too long to a reliance on crude frontal attacks that, even if successful, were wasteful of his limited manpower.

- Lee's opponent at the Battle of Antietam was George McClellan, whose early career had a number of similarities with Lee's. McClellan was born into an old, respected Pennsylvania family and attended West Point, where he was a serious and studious cadet. He specialized in engineering and had served with distinction in the Mexican-American War.

- After the shocking northern defeat at the First Battle of Bull Run, McClellan was summoned to Washington and placed in command of the forces defending the capital. Here, McClellan displayed what would be his greatest talents as a general: training and logistics. He vigorously reorganized the army, enforced modern drill, and greatly improved both its discipline and morale.

- Like Lee, he was popular with the soldiers, who appreciated his constant and energetic efforts to improve their conditions. He also applied his industry to erecting a powerful network of fortifications around Washington that were so extensive and well-made that the Confederates would never make a serious effort to break through them.

- Yet McClellan suffered substantial weaknesses as a general: He persisted in vastly overestimating the numbers of his foes, which led him to be overly cautious and slow to attack. He repeatedly failed to exploit opportunities on the battlefield, keeping large contingents of his troops in reserve when they would have been better employed in attacking the enemy.

- These deficiencies caused him to be relieved of command, but when his successor was thrashed at the Second Battle of Bull Run, Lincoln turned to McClellan to rebuild the broken army. Thus, when Lee followed up the victory at Second Bull Run with his invasion of Maryland, it fell to McClellan to oppose him.

The Battle

- At the outset, McClellan fell into a rare opportunity. Lee had written up his plans for the campaign, including detailed instructions and timetables for the movements of each segment of his army—four groups that would converge on a Union garrison at Harpers Ferry—in a document labeled "Special Order 191." Copies of the order were sent to each of his principal commanders, but one was found lying on the ground by a group of Union soldiers and quickly made its way up the chain of command to McClellan.

- The value of this intelligence was undermined by McClellan's habitual caution. He squandered his opportunity by deploying his army too sluggishly to catch the elements of Lee's army and destroy them, with the result that Harpers Ferry fell to Stonewall

Jackson, and Lee was able to consolidate most of his scattered army near the town of Sharpsburg. He drew up his forces in a defensive line and prepared to give battle.

© Kurt Holter/iStock/Thinkstock.

It is estimated that possession of the center of the battlefield at Antietam switched hands 15 times, and the area became a meat grinder within which entire regiments could be obliterated in an instant.

- McClellan devised a complicated plan of attack that required close cooperation and timing among the various units in order to maintain continuous pressure, but his orders were too specific, and he failed to apprise commanders of the larger picture, so that units acted independently instead of in concert.

- The battle began at dawn on September 17, with elements of the Union right moving forward against the Confederate left, under the command of Stonewall Jackson. The fighting centered on a field of tall standing corn, across which the southern line stretched. Attackers and defenders exchanged deadly fire at close range, and the field was swept by the artillery of both sides, resulting in a bloodbath that shocked even the veterans.

- The momentum of the battle shifted to the center, where the southerners' defenses were anchored by a road that formed a natural trench from which the Confederates could fire on the advancing Union regiments.
    - Lee skillfully fed his limited reserves into the fray at just the right moments to save his formations from collapse. When these were used up, he daringly weakened parts of his line to rush troops to the threatened segments.

- The Union assaults were just uncoordinated and sporadic enough to enable Lee to play this dangerous game successfully, shuffling his outnumbered men in the nick of time to blunt one attack after another.

- Around the road, nearly 6,000 men lay dead or wounded, and the Confederate forces were severely depleted. The formations were shattered, and the defense was reduced to clusters of men still clinging stubbornly to bits of ground. Not only were the Confederates running low on men, but ammunition was scarce, as well.

- McClellan had a body of 14,000 men waiting less than a mile from the road who easily could have been ordered forward to exploit the vulnerable Confederate center. As always, however, he was fearful of imaginary Confederate legions and instead gave orders to the generals commanding his reserves to hold their positions.

- Nevertheless, by early afternoon, the irresistible pressure of the Union attacks had pushed the Confederates back from their initial positions in the north and the center. Although forced to give ground, the southern lines had not broken, and the failure to commit reserves caused the Union advance in these parts of the battlefield to lose momentum. The focal point of the battle now shifted to the Union left.

- The Union commander in this sector was General Ambrose Burnside. His 12,000 men were positioned on one side of a steep-banked stream spanned by a stone bridge. The Confederate units facing him were well-situated on high ground overlooking the bridge in a network of trenches and rifle pits fronted by fallen trees and other obstacles. By midafternoon, Burnside had crossed the narrow bridge and was pursuing the Confederates.

- Lee had nothing left to counter this fresh tide of attackers, and the battle was once again on the verge of a Union victory. But at this crucial moment, the last remaining major division of Lee's army arrived on the battlefield—General A. P. Hill's Light Division.

Hill's division struck the exposed flank of the Union left just as it moved forward against Lee's main line.

- Despite their numerical advantage on the left, it was the Union forces who gave ground and retreated to their initial positions. Hill's exhausted men could not pursue, and the long day of fighting finally ended. The dramatic and timely arrival of Hill had saved Lee's army from destruction.

- The battle of Antietam was the single bloodiest day in the entire Civil War. Stunned by the carnage, the next day, the two armies did nothing, and that night, Lee began withdrawing his forces back toward Virginia. Despite his substantial numerical superiority, his unused reserves, and yet another contingent of 14,000 fresh troops, McClellan let Lee go unmolested.

Outcomes

- Who won the Battle of Antietam? In tactical terms, Lee certainly outmaneuvered McClellan, but in the end, Lee was the one who had to retreat from the battlefield. It is only in retrospect that the true importance of the battle emerges as perhaps the decisive turning point of the war.

- Antietam broke the South's string of victories and gave the Union army confidence that it could stand up to Lee. It solidified Lincoln's position and ensured his party's control of Congress, thus guaranteeing the continued vigorous prosecution of the war. It also ended European flirtation with granting political recognition to the Confederacy, which on its own may have doomed the southern cause.

- Finally, Antietam provided the victory that Lincoln had been waiting for to issue the Emancipation Proclamation. Although mainly symbolic, this act fundamentally transformed the nature of the war from a struggle about states' rights or economics into a moral crusade being fought for the very soul of the nation.

Suggested Reading

Frassanito, *Antietam*.

McPherson, *Crossroads of Freedom*.

Sears, *Landscape Turned Red*.

Questions to Consider

1. Ultimately, how much effect do you think the "lost orders" had on the broader course of history?

2. Do you think Lee could have won the Battle of Antietam, or was a draw the best he could hope for under the circumstances?

1866 Königgrätz—Bismarck Molds Germany

Lecture 30

In the mid-19th century, Otto von Bismarck of Prussia sought to unite under Prussian leadership the dozens of disparate German-speaking principalities and kingdoms into one Germany, which would then become the dominant force in European politics. Bismarck believed that his goals of German unification and the ascendance of Prussia as a great European power would have to be accomplished through military force. Although in retrospect it is tempting to see the unification of Germany as inevitable, a key moment early in the process catapulted both Bismarck and Prussia into prominence: the Battle of Königgrätz, fought against Austria in 1866. Today, this battle is viewed as a crushing Prussian victory and the foundation block of German unification, but it easily could have gone the other way.

The Opponents

- Over the course of 30 years, first as minister of Prussia and then as chancellor of Germany, Otto von Bismarck skillfully pursued the goals of German unification and Prussian ascendancy. Although a brilliant politician, he needed a soldier to carry out his aggressive policies, and in Helmuth von Moltke, he found a partner as skilled at warfare as he was at politics.

Some have argued that World Wars I and II were consequences of the creation of Bismarck's Germany; thus, in achieving his dreams, Bismarck may have created a nightmare for others.

- Moltke was an organizational genius and a brilliant military theorist who oversaw the development of the Prussian army into the most formidable force on the continent. He rose quickly through the ranks, becoming chief of the Prussian General Staff. It was from this position that he was able to put into effect his theories on warfare.

- As a strategist, Moltke managed to combine a gift for meticulous planning and ferocious organization with a pragmatic appreciation for the value of flexibility in real warfare.
  - He favored movement and flank attacks over fixed fortifications, even when on the defensive.
  - He was one of the first to realize that the growth in firepower on the battlefield was rendering frontal assaults increasingly costly and outdated.
  - He was ahead of his time in appreciating the value of a well-trained general staff and was a proponent of new technologies, making extensive use of railroads to mobilize his troops during the war with Austria and of the telegraph to coordinate the movements of widely dispersed armies.

- Moltke was not able to fully modernize Prussia's artillery, but he was successful in organizing and equipping its infantry along state-of-the-art principles, particularly the basic gun carried by the men. Whereas their Austrian opponents were still using muzzle-loading muskets, Prussian infantry were issued modern bolt-action, breech-loading rifles.
  - Although the breech-loader offered a number of advantages, initially, there had been difficulty in making a breech strong enough to stand up to the force of repeatedly firing the weapon.
  - By the mid-19th century, some arms manufacturers were solving that problem. One version used a sharp, needlelike spike to strike the cartridge and fire the weapon. Because of its firing mechanism, it was often popularly called a needle gun. This gun was selected by the Prussian army for mass production and issuance to its troops.

- During this period, the two most powerful states on the continent were France and the Austrian Empire. A united Germany would challenge their hegemony; thus, any attempt to create a powerful new nation would likely involve direct conflict with them. Austria, in particular, exercised varying degrees of control over several of the key Germanic states and regions that Bismarck wished to bring within the Prussian sphere of influence.

- The Austrian Empire was a large, wealthy, culturally fragmented realm that included parts of northern Italy, Hungary, Croatia, the Illyrian coast, Bohemia, and Transylvania. The Austrian army, especially its fine Magyar cavalry, enjoyed a fine reputation, having fought with success in a number of minor wars.

- Commanding the Austrian forces at Königgrätz was Ludwig von Benedek, the son of a Hungarian doctor. Like Moltke, he determined upon a military career early in life and enrolled as a cadet in a military academy at the age of 14. He was both a highly popular and a successful officer, fighting in a number of conflicts, mostly in Italy, and gaining a reputation for bravery. At the Battle of Solferino in 1859, when the other elements of the army fled, his command alone stood its ground and nearly snatched victory from disaster, further enhancing his reputation.

- When war with Prussia became imminent, Benedek was the popular choice for overall command of Austria's armies, but he was reluctant to accept, recognizing that grand strategic thinking was not his strength. Among other factors, he had spent his entire career fighting in Italy and, thus, was unfamiliar with even the basic topography of Central Europe. Nevertheless, he accepted the post and filled his staff with similarly ill-suited aides. The contrast between the slapdash, rather disinterested General Army Staff of Austria and Moltke's highly efficient, machine-like Prussian one could not have been more pronounced.

The Campaign and Battle

- Bismarck and the Prussians knew that to supplant the Austrian Empire in its dominant position over the German confederation, they would have to provoke a war. The opportunity came when Prussia came into conflict with Austria over Schleswig-Holstein.

- On June 15, 1866, the Prussians demanded that the states of Hannover, Hesse-Kassel, and Saxony abandon their alliances with Austria and disarm. When they refused, the war officially began. Now, Moltke's careful planning paid dividends, as Prussian armies rapidly mobilized and swept forward, and the three states were conquered with astonishing swiftness.

- The subjugation of these states was the first step in an overall plan for invading the Austrian Empire. Moltke had devised a four-pronged attack, in which the separate armies would conquer initial objectives, mostly the minor German states, then combine to take on the main Austrian army.

- Critics of Moltke's plan feared that dispersing the elements of the Prussian army so widely would make them vulnerable to attack, but the speed of their advance kept the Austrians off balance, and the invasion accomplished all of its initial objectives.

- Meanwhile, Benedek situated the main Austrian army in an arc with its back to the Elbe River northeast of Königgrätz in southern Bohemia. This Austrian army was as large as all the Prussian ones combined, numbering around 250,000 men.

- By dawn on the morning of July 3, the Prussian Elbe and First Armies were closing in on the Austrian army from the east and were ready to launch an attack. The Second Army, however, was still marching down from the north. Without it, the Elbe and First Armies were outnumbered almost two to one. Moltke had to decide whether or not to launch the attack with the available forces and hope that the Second Army would arrive to strike the Austrians' exposed northern flank. He decided to attack.

- On the Prussians' right flank, the Elbe army managed to cross a key bridge but encountered determined resistance. In the center and on the left, the Prussian First Army waded across a river and pressed its attack, initially meeting with success and driving back the opposing forces. The needle gun proved its superiority and helped the Prussians to defeat larger contingents of Austrians.

- The Prussians advanced through a series of small villages before entering a zone commanded by the massed Austrian artillery. The Austrian rifled cannons were of excellent quality, and more than 250 guns unleashed a firestorm that pounded the Prussian ranks, inflicting heavy casualties.

- On the Prussian left, many men sought refuge from the deadly cannonade in the Svir forest, which became the scene of an intense fight. The Austrians launched a counterattack that halted the advance and threatened to break the Prussian lines. At least 13 Austrian charges were flung against them, and they were on the verge of buckling. The desperate Prussian commander appealed for reinforcements, but Moltke refused, calculating that until the Second Army arrived, all reserves had to be kept intact to meet the potential threat of a general Austrian advance.

- One Austrian commander pushed his men forward on the far left of the Prussian line and was in position to outflank them. All he needed was for Benedek to support this action with a push against the Prussian center. This was a crisis point in the battle and the moment of supreme danger for the Prussians. Benedek, however, chose to remain on the defensive, refusing all requests to advance and, instead, letting his cannons chew up the Prussian forces.

- Moltke, aware of the danger, knew that all he could do was watch for the Second Army. At last, it appeared and began to drive into the Austrians' right flank. At around the same time, the Elbe army finally broke its stalemate and enveloped the left flank of the Austrians. Caught in a classic double envelopment, the Austrian army began to crumble.

- Although some of the Prussian generals wanted to pursue and completely destroy the Austrian army, Bismarck astutely realized that, in the long run, it would be more useful to reconcile with Austria rather than to have her as a bitter enemy; thus, he persuaded Moltke to call off the pursuit.

Outcomes

- After the battle, the Austrian Emperor Franz Josef immediately sued for peace. Austria was finished as a great power and, within six months, was reorganized as the Austro-Hungarian Empire.

- In 1870, with Moltke in command, Prussia inflicted a swift and humiliating defeat on the French. As part of the peace treaty, France was forced to cede Alsace-Lorraine and pay a huge indemnity—actions that created great resentment and were factors leading to the outbreak of the First World War.

- The only European nation more powerful than a united Germany was Great Britain, and the two countries engaged in an arms race that was yet another element leading to World War I. It is not difficult to see how the political and military history of the first half of the 20th century can be traced directly to Prussia's triumph at Königgrätz, but it should not be forgotten how easily the battle might have had a different outcome.

Suggested Reading

Craig, *The Battle of Königgrätz.*

Steinberg, *Bismarck.*

Wawro, *The Austro-Prussian War.*

Questions to Consider

1. Who do you think contributed more to German unification, Bismarck or Moltke?

2. What were the pivotal factors that determined the outcome of the Battle of Königgrätz?

1905 Tsushima—Japan Humiliates Russia
Lecture 31

On July 8, 1853, the farmers and fishermen living around the port of Uraga, Japan, witnessed a disturbing sight: Four foreign warships boldly steamed into the bay and anchored just offshore; they were commanded by Commodore Matthew Perry, an American naval hero. For the previous 250 years, Japan had pursued a strict isolationist policy, closing its borders to almost all contact with foreigners and refusing even to meet with representatives of other nations. This policy had resulted in Japan missing the Industrial Revolution. At least 18 attempts by various countries had sought to establish economic or diplomatic ties with Japan, but every one of them had been rebuffed. Perry was determined to succeed where others had failed.

Japan's Military Modernization

- In July of 1853, through a combination of blunt force, bullying, and stubbornness, Commodore Matthew Perry, a U.S. naval hero, managed to make contact with representatives of the Japanese government, penetrating a strict isolationist policy that had been in place for the previous 250 years. On a follow-up visit the next year, he negotiated the Treaty of Kanagawa, which opened Japanese ports to American trade.

- Much of Perry's success was due to Japan's lack of a real navy. This military inadequacy was intensely humiliating for the Japanese, and it was one of the main factors that sparked the Meiji Restoration. The new leadership adopted an astonishingly aggressive policy of modernization and industrialization.

- The Japanese decided to mold their new army and navy on the best foreign models they could find.
  - Great Britain had the reputation of having the largest, most efficient, most well-trained, and most technologically advanced navy in the world. Accordingly, Japanese naval

officers and engineers were sent to Britain to study the methods and practices of the Royal Navy, and contracts were signed for new Japanese warships to be constructed in British shipyards following the most up-to-date designs.

- o The modern Japanese army was initially to be based on the French model, but after Prussia's impressive defeats of Austria in 1866 and France in 1870, the Japanese switched to the Prussian military, again dispatching observers and purchasing the latest in German rifles and cannons.

- Nevertheless, Japan continued to be viewed as a strange but unthreatening and technologically backward nation. This assumption would be blown to bits in the narrow gap of water separating Japan from Korea, the Tsushima Strait. It was here that, on May 27, 1905, a thoroughly modern Japanese fleet annihilated a much larger Russian one, announcing Japan's entrance onto the global stage as a major power.

The Navies and Their Technology

- At the outbreak of the Russo-Japanese War, the Japanese fleet had 6 modern battleships, all built in Great Britain within the previous seven years. The flagship was the *Mikasa*, a 15,200-ton battleship completed in 1902. It represented the state of the art, having 12-inch main guns that could be reloaded in whichever direction the guns pointed, a new technology that increased its rate of fire.

- Next in size and strength were 8 powerful armored cruisers, also recently built in foreign shipyards and all well-armed with 8- and 6-inch guns. These 14 ships formed the core of the battle fleet, supported by several dozen smaller cruisers and destroyers, as well as some squadrons of torpedo boats.

- The Japanese sailors were highly disciplined and dedicated. Because they had been subjected to vigorous training and drills, they knew their roles and had experience using their ships and weapons.

The armament of the Japanese ship Mikasa included four 12-inch cannons in two rotating turrets, as well as a large secondary armament of 6-inch guns.

- The admiral in charge of the Japanese fleet was Heihachiro Togo. As a young man, he witnessed an incident in which a British warship was able to shell Japan with impunity because of its lack of an effective navy, causing him to pursue a naval career. Accordingly, he was sent to England as a cadet in the British navy. After seven years, he returned to Japan, serving aboard a variety of warships and participating in several minor sea battles. In 1895, he attained the rank of admiral.

- Russia had a much larger fleet divided into three groups: the Baltic Squadron, which had the most modern battleships; the Black Sea Squadron, trapped in that body of water by Turkish control of the Bosphorus; and the Pacific Squadron. The Pacific Squadron, based near Japan at Port Arthur, consisted of 7 battleships, 7 cruisers, and a smattering of smaller craft. Although some of the vessels were reasonably modern, others were outdated, and the crews were generally ill-trained.

The Campaign

- Tsar Nicholas II of Russia had determined to focus his expansion efforts in the east and began to commit more troops and ships to increasing Russian power there. The obvious targets were Manchuria and the Korean peninsula. But Japan, whose imperial ambitions were growing along with its modernized military, also coveted these two regions. On February 4, 1904, Japan severed relations with Russia, and the Russo-Japanese War officially began.

- The Japanese struck first, with Togo launching a surprise torpedo attack against the Russian ships anchored at Port Arthur. Psychologically, the attack was a great blow to Russian confidence and an equally powerful boost to Japanese morale. The tsar dispatched the most renowned admiral in Russia, Stepan Makarov, to Port Arthur to take control of the situation and restore Russian pride.

- Makarov quickly refloated the battleships and revived the spirits of the Pacific fleet. Unfortunately for the Russians, while leading the fleet on its very first sortie, Makarov's flagship struck a mine and sank within minutes, carrying Makarov and 662 members of its crew to the bottom.

- Togo now had the upper hand. The Japanese launched a massive invasion, ferrying hundreds of thousands of troops to Manchuria, defeating Russian armies at the battles of the Yalu River and Nanshan, capturing the port of Dalny, and surrounding Port Arthur.

- The tsar, determined to make a maximum effort to win the war, sent the Baltic Squadron to reinforce the Pacific Squadron. Renamed the Second Pacific Squadron, this was a powerful force that included 4 new battleships, 3 older battleships, and an attendant swarm of cruisers, destroyers, and support ships.

- The man selected to lead the Second Pacific Squadron was Admiral Zinovy Rozhestvensky. He was a strict disciplinarian, but he was also fair-minded and relatively concerned for the welfare of his men. Rozhestvensky spent two years as a naval attaché in Great

Britain, where like Togo, he witnessed firsthand the organization and technology of the Royal Navy.

- Just getting his ships to the battle zone posed an enormous logistical challenge. The vessels were coal-powered, and there were not many friendly refueling stations along the route. The Hamburg–America steamship line was contracted to position supply ships bearing 340,000 tons of coal at intervals along the route. On October 15, 1904, the Second Pacific Squadron departed on what proved to be an epic journey.

- Meanwhile, Port Arthur was being menaced by advancing Japanese troops; thus, the Russian fleet there was ordered to move to the port of Vladivostok. The remaining battleships and cruisers were intercepted by Togo. In the Battle of the Yellow Sea, the more efficient Japanese gunners disabled the Russian flagship and sank 2 cruisers. The rest of the Russian fleet fled to Port Arthur or to neutral ports, where they were interned for the remainder of the war.

- After the destruction of the original Pacific Squadron, Tsar Nicholas decided to reinforce Rozhestvensky's fleet. Nearly anything that could float and had a gun was rounded up and organized as the Third Pacific Squadron. The ships in this group included a number of hopelessly outdated ironclads and slow coastal defense vessels. Rozhestvensky realized that these ships would add nothing to the offensive capability of his fleet and tried to persuade his superiors not to burden him with them, but he was nevertheless ordered to wait for their arrival before proceeding to Vladivostok.

The Battle

- Rozhestvensky's combined fleet, now with nearly 50 vessels, finally approached the vicinity of Japan in late May 1905. Despite the presence of many ships of questionable value, Rozhestvensky still had a solid core of 7 modern battleships to Togo's 4 and the advantage of more 10- and 12-inch cannons—41 to Togo's 17. The Japanese, however, had a significant edge in training, gunnery, speed, morale, and 8-inch guns.

- Early on May 27, a Japanese ship sighted the Russian fleet in the Tsushima Strait and radioed its position and course. Togo moved to intercept, and the two fleets drew near each other. Because their approach would have put Togo's cruisers into action first, rather than his battleships, Togo had his entire fleet execute a semicircular turn within range of the Russian guns, exposing each ship to concentrated fire for a few moments. But the Russian gunners were unable to take advantage of the opportunity, and the maneuver was pulled off without serious damage.

- Both fleets were deployed into several columns, and the battle now began in earnest, with the two sides trading fire at relatively close range. Within 20 minutes, the well-directed gunnery of the Japanese began to take its toll, including the loss of Russia's newest battleship and the severe wounding of Rozhestvensky.

- With the Russians leaderless and having lost their best ships, the battle began to become a massacre. The fighting continued into the night, with the Japanese hunting down the remaining groups of Russian ships. With the destruction of its navy, Russia had to ask for peace talks. Japan gained Port Arthur and Sakhalin Island and was effectively free to do whatever it wanted in Korea.

Outcomes

- In Russia, the disasters of the Russo-Japanese War led to the uprisings of 1905. Although temporarily suppressed, these social tensions continued to boil until finally exploding in 1917 as the Russian Revolution.

- The degree of success Japan experienced had the long-term result of encouraging and strengthening the militaristic and imperialistic elements within Japan. This would lead to continued expansionist policies and territorial aggression in the early 20th century, culminating in another invasion of Manchuria during the 1930s.

- With Russia removed from the equation, the only navy that could stand up to Japan in the Pacific was that of the United States, and the rivalry between the two nations steadily increased until they came into conflict in World War II.

- Thus, the Battle of Tsushima can be said to have directly contributed to both the fall of tsarist Russia and the rise of a militaristic Japan whose expansionist policies contributed to the global cataclysm of World War II.

Suggested Reading

Corbett, *Maritime Operations in the Russo-Japanese War, 1904–1905.*

Jukes, *The Russo-Japanese War*.

Pleshakov, *The Tsar's Last Armada*.

Questions to Consider

1. What important technological advances in naval warfare are illustrated by the Battle of Tsushima?

2. Do you think the Russians ever had a realistic chance of beating the Japanese navy? Why or why not?

1914 Marne—Paris Is Saved

Lecture 32

On September 2, 1914, French Flight Lieutenant Watteau took off on a reconnaissance mission that would change history. In less than a month, the German armies had swept through Belgium and across northern France and were within 40 miles of Paris. Watteau's mission was to scout the movements of the westernmost pincer of the German assault, the army of General Alexander von Kluck. But Watteau noted that instead of continuing their westward push, the German soldiers were marching east. When the French high command received this report, they realized that this turn exposed the flank of Kluck's army to possible attack. The subsequent battle, fought along the Marne River, would result in an allied victory so dramatic that it is often called "the miracle of the Marne."

The Opponents

- By 1914, the nations of Europe were bound together in a complex web of treaties and agreements that virtually ensured that if any two of them went to war, all the rest would be drawn in, as well. Against this diplomatic background, tensions were rising that made the outbreak of conflict increasingly likely.

- Among these tensions were the ongoing naval arms race between Britain and Germany, ethnic unrest in the Balkans, German imperial ambitions, French resentments lingering from the Franco-Prussian War, and territorial rivalries between Austria-Hungary and Russia. As it turned out, the spark that ignited World War I and set into motion the preordained chain of alliances and declarations of war was the assassination of the Austrian Archduke Franz Ferdinand by a Serbian.

- The greatest challenge for the Germans in a general European war was facing France on one side and Russia on the other. In 1894, then Chief of Staff Alfred von Schlieffen created a plan that had Germany immediately launching a massive offensive against

France as soon as war was declared and defeating it within weeks. Then, using Germany's excellent rail system, the troops could be rushed back to the Russian frontier.

- The powerful right wing of the German army would swing to the north, crossing through neutral Belgium before enveloping Paris and the main body of the French army. The violation of Belgian territory would most likely bring Great Britain into the war, but the British army had fewer than 100,000 men available for quick deployment in Europe.

- It was a bold plan, with little room for error or the unexpected. In Schlieffen's concept, everything depended on the right wing, which had to be powerful enough to punch through any opposition and keep advancing without losing momentum. He was willing to risk weakening other sectors to ensure that it would be strong enough and constantly fretted that less daring commanders might alter his plan by distributing the troops more evenly.

- His successor as Chief of Staff was Helmuth von Moltke, the nephew and namesake of the great Moltke who had led Germany to victories in the Austro-Prussian and Franco-Prussian wars. Moltke assigned eight new divisions to the German center and added only one to the crucial right wing. Worse, at a key moment, he lost his nerve and prematurely shifted two corps from the French front to the Russian, further weakening the power of the German offensive thrust.

- The French General Staff made several incorrect assumptions about how the Germans would behave.
    - First, they didn't believe that the Germans would violate Belgian neutrality; thus, they were insufficiently prepared to meet an attack from that direction.

    - Second, they grossly underestimated the number of troops the Germans could field. At the outbreak of war, the French estimated that they were facing 43 German divisions, when there were actually 83.

- Finally, the French had invested in building a chain of fortresses along the border with Germany and were convinced that any attacks would be aimed at these and that such attacks would be doomed against their well-protected batteries of heavy cannons.

- In the years just before the war, the French had begun to shift from a defensive mindset to an offensive one, as reflected in their plan for the outbreak of war—Plan 17. They had become convinced that wars were decided by morale and that the side with the greater will to conquer would be victorious. The official field regulations of the French army were rewritten to reflect this philosophy. Thus, at the outbreak of war, Plan 17 called for four French armies to launch a powerful attack on Alsace and Lorraine.

The Campaign

- On August 3, 1914, Germany declared war on France, and all these long-formulated plans were put into motion. Every day, 550 well-organized trains transported German soldiers and equipment across the Rhine, launching nearly a million and a half men against France. By August 4, German troops were pouring into Belgium, and Great Britain declared war on Germany.

- The crucial right wing of the German attack was the First Army, commanded by General Alexander von Kluck. The plan ran into its first obstacle when the Belgians did not submit but instead fought back. Several Belgian fortresses put up spirited defenses and managed to hold out until August 15. Despite this setback, the Germans were close to maintaining their original schedule

- Meanwhile, the French had put Plan 17 into motion. With the German defenders proving to be both more numerous and more deadly than French plans had allowed for—in less than a week, more than 300,000 French soldiers were mowed down—the offensive bogged down in the forests. By August 24, the French armies were forced to pull back to their original lines.

- While Plan 17 was foundering against the German center, the British Expeditionary Force (BEF), a highly professional army of about 80,000 men, had been deployed near Mons. Here, they fought a sharp engagement against Kluck and briefly checked his advance, but they were in an exposed position and had to retreat.

- In this moment of crisis, the overall commander of the French forces opposing Moltke was General Joseph Joffre. He was supremely confident and utterly unflappable, and in this particular crisis, he was exactly the right man for the job. He coolly drew together the shattered remnants of the French armies and began methodically assembling a new Sixth Army to defend Paris.

The Battle

- The unexpected Belgian resistance, the enthusiastic French attacks on the frontier, and the strong showing of the BEF undermined Moltke's confidence in the plan, and he miscalculated the strength of his opponents. Now, from the eastern front came news that the Russians had defied German expectations by mobilizing enough men to invade East Prussia.

- On August 25, Moltke ordered two army corps to be withdrawn from the assault on France and sent to the eastern front. By the time they arrived, the Russians had already suffered a horrific defeat. In retrospect, the redeployment of the two army corps meant that at the very moment when the German forces in the west needed to press home their attack, they found themselves with a sizable gap in their lines and a significant part of their striking force gone.

- The apparent weakness of the French opposition caused Moltke to consider deviating from the initial plan for enveloping the French armies and to try breaking straight through them. He issued orders altering the role of Kluck's army from the main enveloping force to supporting the flank of the more central armies. At first, Kluck continued his drive west, but he, too, was worried that his weakened army might not be able to complete its wide sweep. Thus, around

September 1, his army began the fateful change in direction detected by the reconnaissance aircraft of the allies.

- On August 26, General Joseph Gallieni was appointed military governor of Paris and charged with its defense. He energetically threw himself into this role and was one of the first to recognize the opportunity that the shift in Kluck's direction offered for attacking the flank of the enemy. Once the change of direction had been confirmed, Joffre determined to use the new Sixth Army.

- Joffre gained British cooperation, and by September 5, the counterattack was underway. Kluck began to pivot to try to meet the danger, but this maneuver had the effect of creating a dangerous gap between his army and the adjacent Second German Army. The confrontation took place near the Marne River, with the first serious fighting near the village of Saint-Soupplets.

© JOHN GOMEZ/iStock/Thinkstock.

The initial trenches dug after the Battle of the Marne would proliferate, transforming the landscape of northern France into a grotesque world of mud, barbed wire, and shell holes.

- Over several days, the German advance was fought to a standstill, and the counterattacking French elements began to arrive and make their presence felt on the German flank. The BEF inserted itself into one of the gaps between the German armies, where it threatened to cut off Kluck's army. On September 9, the commander of the German Second Army recognized that the offensive had come up short and made the decision to start withdrawing his troops. Kluck's First Army soon followed. Paris—and France—were saved.

Outcomes

- The halting of the German advance at the Battle of the Marne left both sides enmeshed in a war that neither had anticipated or prepared for. With the failure of the Schlieffen Plan, the Germans had lost their best chance to win the war, and by failing to aggressively follow up the victory at the Marne, the allies had probably lost their best opportunity to bring the war to a rapid end.

- Each side was now locked into an escalating and seemingly interminable bloody stalemate. By the time it was finally over, 10 million men would be dead and another 20 million wounded. It could be argued that all the horrible and destructive remainder of World War I was the legacy of the Battle of the Marne and the chances that both sides missed to end the war quickly.

Suggested Reading

Herwig, *The Marne.*

Sumner, *The First Battle of the Marne.*

Tuchman, *The Guns of August.*

Questions to Consider

1. If the Germans had stuck to the spirit of Schlieffen's original plan, could they have won World War I?

2. How did the technological advances in military equipment in the early 20th century lead to the stalemate of trench warfare, and was this inevitable?

1939 Khalkhin Gol—Sowing the Seeds of WWII
Lecture 33

The Battle of Khalkhin Gol, fought between Russia and Japan in 1939, was an odd conflict on nearly every level. It arose out of a border dispute concerning a 10-mile strip of featureless steppe at the point where Mongolia, Manchuria, and China converge. It was an accidental encounter that neither side sought, yet it ultimately involved fighting that lasted for five months and resulted in tens of thousands of casualties—and it wasn't part of any officially declared war. Khalkhin Gol was an important technological testing ground for new weapons, yet it remains a relatively little-known event. Probably the most surprising aspect of its obscurity is that Khalkhin Gol exerted a number of direct and significant effects on World War II.

Background to Khalkhin Gol

- Since Japan embarked on its rapid industrialization in the late 19th century, its greatest challenge had been obtaining the natural resources required by large-scale mechanization. Thus, Japan's imperial ambitions focused on conquering regions where these raw materials could be found. Logically enough, Japan first looked to the mainland of Asia.

- The military commanders were the most aggressive advocates for expansionism in Manchuria, Korea, and other mainland regions, and because the military enjoyed the emperor's favor, this policy prevailed, resulting in Japan's occupation of Manchuria and China in the 1930s. In these campaigns, the Japanese army experienced a string of successes that had made them overconfident, and they now began to eye Mongolia and Siberia.

- During the 1930s, the portion of the Japanese armed forces that occupied Manchuria, now organized into the puppet state of Manchukuo, had developed an odd structural relationship to the rest of the military. It was known as the Kwantung Army after the province where it originated, and it operated semi-autonomously, although its commanders were expected to follow general Japanese policy.

- The immediate cause of the Battle of Khalkhin Gol was the ambiguous borders between the Japanese puppet state of Manchukuo and the Russian puppet state of Mongolia. In particular, the Japanese maintained that for a stretch of about 30 miles, the Halha River—called the Khalkhin Gol by the Mongolians—demarcated the boundary. The Russians instead claimed that the border ran about 10 miles to the north.

The Battle

- In March 1939, Major Masanobu Tsuji of the Kwantung army staff drafted a new set of guidelines for how to respond to minor border clashes. Major Tsuji was a fanatical Japanese militarist and nationalist who viewed foreigners as racial inferiors and believed that any sort of brutality was justified in dealing with them. His set of instructions decreed, "If any Soviets cross the frontiers, annihilate them without delay," and "Where boundary lines are not clearly defined, local commanders will upon their own initiative establish boundaries."

- On May 11, 1939, a 20-man patrol of the People's Republic of Mongolia was spotted moving through the disputed zone between the Halha River and the village of Nomonhan. A group of 40 Manchukuan cavalrymen drove them away after a brief skirmish. Both sides then reported the incident. The Japanese sent more than 50 aircraft to search for the enemy.

- Meanwhile, the Russians also responded, ordering a battalion of infantry to the scene, accompanied by some light tanks and artillery. When the Japanese detected this force, they decided to implement Tsuji's directive to annihilate the enemy and dispatched more than 2,000 men, supported by artillery and tankettes, to locate and crush them.

- A pitched battle took place on May 28. The Japanese had badly underestimated both the size and the quality of the Russian opposition. The Russians had equipped even their light tanks and armored cars with relatively heavy cannons, which chewed up the weakly armed and armored Japanese.

- The Japanese high command wanted to avoid this distraction, but Tsuji argued passionately that a failure to act firmly would invite invasion. The Kwantung Army staff eventually came around and committed a large force to a renewed attack. This group featured Japan's only independent tank brigade, which boasted two regiments of medium tanks and one of light tanks. The Japanese estimated that to oppose this powerful army, the Russians could muster only around 1,000 infantry and a dozen tanks.

- However, undetected by the overconfident Japanese, the Russians had reinforced the area with a large army of their own. Most important, against the 70 rather flimsy Japanese tanks, the Russians had deployed 186 tanks and 266 armored cars, almost all of them more heavily armed. In addition, a new commander had been appointed. He was Georgy Zhukov, and he would prove to be one of Russia's greatest generals.

- After some preliminary aerial warfare, the Japanese launched their offensive, executing a successful night attack in which their tanks charged and scattered an unsuspecting Russian infantry regiment. Zhukov responded with a counterattack by his tanks, but they were unable to coordinate their movements with the supporting infantry, and many of his superior armored vehicles were knocked out by Japanese suicide attacks.

- Zhukov was learning to coordinate his tanks, artillery, and planes, and his counterattacks first stopped the Japanese advance, then began to wipe out the surrounded pockets of Japanese troops and vehicles. The remaining Japanese had no choice but to attempt to retreat across the river under heavy fire. There were further

Japanese assaults, but the momentum had decisively shifted in favor of the Russians.

- With the Japanese offensives thwarted, Zhukov set about gathering strength to launch his own major attack. He devised a plan in which his center would hold the Japanese in place while his tank-heavy right and left wings would sweep around and encircle the enemy. On August 20, he launched his attack, and by August 29, the Japanese were trapped in three pockets and largely wiped out.

Outcomes

- This undeclared border war had important and far-reaching effects on both Japan and Russia, as well as on the course and outcome of World War II. For the Japanese, the shocking failure of their army at Khalkhin Gol resulted in a dramatic reorientation of strategic plans.
    - Before the battle, the army had been dominant and favored a policy of northern expansion. Now the Japanese navy gained the upper hand. Rather than advocating attacking on the Asian mainland, the navy urged a policy of southern expansion, in which the targets for Japanese imperialism would include the resource-rich regions of the Dutch East Indies, French Indochina, and the Philippines.

    - The only force that could constitute a threat to this southern-directed imperialism was the U.S. Navy. Therefore, Japanese war planning concentrated on eliminating the U.S. Navy to enable the Japanese to carry out their planned conquests. The solution was an attack on Pearl Harbor.

- On the Russian side, the battle also produced momentous consequences. During the summer of 1939, both Britain and Germany were actively seeking alliances with Russia. On the one hand, the British and the French desired Russian aid in curbing Hitler's ambitions; on the other, Hitler felt that he could not launch his planned invasion of Western Europe until he had ensured that the Russians would not attack him.

- In the end, Stalin chose to sign a nonaggression treaty with Germany that included secret provisions carving up Poland between them and granting each side "spheres of influence" in the Baltic. One week after the agreement was signed, German tanks rolled into Poland, officially beginning World War II.

- A main reason why Stalin choose to side with Hitler, especially when there were signs that Hitler would eventually turn against Russia, was the Khalkhin Gol conflict, which was being fought while Stalin and Hitler were negotiating. Stalin viewed Japan as a perennial threat on his eastern frontier and feared a two-front war against Japan in the east and Germany in the west. Khalkhin Gol therefore played a key role in initiating World War II in Western Europe.

- Khalkhin Gol also influenced the course of the war. Stalin's paranoia had caused him to undertake a sweeping purge of army officers that stripped the Russian military of almost all its experienced leaders on the eve of World War II. When Hitler finally broke the nonaggression treaty and invaded Russia in June 1941, these purges had left the Russian army ill-prepared to resist.

- Yet a core of men and officers with battle experience did exist and were dispatched to the western front, where they had a major part in stopping Hitler's invading armies: the troops who had fought at Khalkhin Gol. Foremost was General Zhukov himself. Because of his success at Khalkhin Gol, he was appointed Chief of the General Staff, and in that role, he applied his experience and knowledge of integrated massed infantry, tanks, artillery, and aircraft to rally the Russians and fight the German advance to a standstill.

- The eastern troops and equipment were essential, as well. Khalkhin Gol provided a testing ground for many new weapons, particularly tanks. At Khalkhin Gol, the Russians were able to try out some of their new designs for armored vehicles and to learn to install in their tanks large cannon and hatches that could be locked from the inside. When the Germans invaded, the nastiest surprise was the very high quality of the Russian tanks, which in many cases, were better armed and armored than the German panzers.

- The experience Russia's troops gained in the east also paid dividends. At the moment when the German invasion posed its greatest threat, the Russians were able to shift 18 army divisions, 1,700 tanks, and 1,500 aircraft from Mongolia to the western front, and the infusion of these experienced soldiers was critical to stopping the German advance.

- Zhukov oversaw the pivotal victory at Stalingrad and directed the Russian counterattacks against the Germans that drove them out of the Soviet Union and back into Germany. In these offensives, Zhukov again made use of encirclement and massed tank, artillery, and air attacks that he had perfected at Khalkhin Gol. Hitler's invasion came close to succeeding; if not for Khalkhin Gol, Russia might well have lacked the leadership, the tactics, the technology, and the soldiers that ultimately made the difference.

Suggested Reading

Coox, *Nomonhan: Japan against Russia.*

Goldman, *Nomonham 1939.*

Questions to Consider

1. What differing approaches to warfare were displayed by the Japanese and the Russians in the Battle of Khalkhin Gol?

2. Do you agree or disagree with the arguments made for the long-term effects of Khalkin Gol? Why or why not?

1942 Midway—Four Minutes Change Everything

Lecture 34

In the 35 years following the Battle of Tsushima, big-gun battleships had ruled the waves, but on December 7, 1941, the Japanese attack on Pearl Harbor signaled the dawn of a new era of naval warfare: The once-fearsome battleship was obsolete, and airplanes were now the dominant weapon. Indeed, the outcome of the war in the Pacific, which both sides had anticipated would feature close-range slugging matches between rows of battleships, would instead be decided by dueling aircraft carriers, separated by hundreds of miles, flinging squadrons of bombers and torpedo planes at one another. The most dramatic and pivotal of these carrier-versus-carrier matches took place in June of 1942 around the island of Midway.

The Opponents

- The person who most shaped the Battle of Midway was the commander of the Japanese fleet, Isoroku Yamamoto. Yamamoto studied at the Japanese Staff College and was then sent to the United States for several years. He attended Harvard University and was appointed as a naval attaché in Washington, D.C. Back in Japan, he commanded several ships, including the aircraft carrier *Akagi,* and this experience made him an early advocate of air power in naval warfare.

- Yamamoto was part of the naval faction that resisted the army's adventurism in China and Manchuria, and he opposed provoking a war with the United States in the Pacific, fearing that the Americans would be able to use their enormous natural resources and industrial capacity to overwhelm Japan. When Japan nevertheless committed itself to a policy of expansion in the Pacific that would inevitably bring it into conflict with the United States, Yamamoto had to devise a plan to maximize Japan's chances of victory.

- His solution was a surprise attack on the U.S. fleet that would destroy or disable its battleships and aircraft carriers, freeing the

Japanese navy to capture the Philippines and the Dutch East Indies, whose natural resources were vital raw materials for Japan's military. The raid on Pearl Harbor fulfilled his expectations.

- What did not go according to plan was that the U.S. aircraft carriers were out of port when Pearl Harbor was attacked. As the war progressed and it became clear that victory at sea would be determined by air power, all that stood between Japan and its goal of achieving total domination over the Pacific was this handful of American aircraft carriers. Yamamoto began to craft a plan to lure these ships out to do battle where the superior Japanese forces could catch and sink them.

- Yamamoto had always had a predilection for complex, multipart operations, but his scheme to trap the American carriers was the most elaborate yet. It featured a vast armada and thousands of soldiers on troop transports to carry out planned invasions. The main strike force of four large carriers was commanded by Admiral Nagumo.

- The key to the plan was the tiny island of Midway, the last outpost in the American defenses before Pearl Harbor. The main Japanese strike force would bomb Midway, and troops would land and occupy it. Meanwhile, another task force would invade the Aleutian island chain leading to Alaska. Yamamoto was sure that the Americans would send their aircraft carriers and the remnants of the U.S. Pacific Fleet from Pearl Harbor to defend Midway. There, they would be attacked by planes from the Japanese carriers and sent to the bottom of the ocean.

- The Japanese believed that the Americans could not muster more than two carriers to oppose them at Midway; thus, the four Japanese carriers would be more than enough to deal with the Americans. In reality, they would face three large American carriers that could launch 233 aircraft, compared to the 248 that the four Japanese carriers could launch. Despite the size of their armada, Japanese overconfidence allowed a lopsided contest to even out in the category that really mattered.

- The Japanese were unaware that the Americans had advance warning of Yamamoto's plan. American intelligence agents had managed to partially decipher their code; thus, the American commanders knew the basic outline of the Japanese strategy, including the fact that Midway was the target and when the Japanese intended to strike it.

- Yamamoto's American counterpart was Admiral Chester Nimitz, a tough, no-nonsense leader with a long career in the navy. In the aftermath of Pearl Harbor, he was appointed commander-in-chief of the Pacific fleet. At Midway, his chosen delegates to conduct the battle were the highly experienced carrier commander Jack Fletcher and Raymond Spruance, known for his calm demeanor and calculating mind.

- Once he knew Yamamoto's plan, Nimitz saw an opportunity to turn the tables and potentially deliver a serious blow. He heavily reinforced the island, especially the airfield, stationing a squadron of B-17 bombers there. He positioned his carriers 325 miles northeast of the island, where he hoped they could ambush Yamamoto's carriers.

The Battle

- At dawn on June 4, the main Japanese carrier group launched a strike force of 108 planes against Midway. Fighter planes from Midway intercepted it, and the outdated and outclassed American planes were shot down by the faster, more maneuverable, more heavily armed Japanese. Midway's airfield was bombed.

- Meanwhile, both sides had been deploying reconnaissance aircraft to search for the other. Most of Nagumo's reconnaissance planes were launched by catapults from his heavy cruisers. One, Number 4, was half an hour late taking off from the cruiser *Tone* because of problems with its catapult. As chance would have it, the sector of ocean that this plane had been assigned to search was the one

© Getty Images/Photos.com/Thinkstock.

The Japanese raid on Pearl Harbor destroyed or disabled the U.S. battleship fleet but also marked the beginning of an era in which airplanes become the dominant weapon in naval warfare.

containing the American carriers; thus, Number 4's sighting of them was delayed a crucial half-hour.

- Meanwhile, Fletcher and Spruance had already received reports locating at least two Japanese aircraft carriers. Beginning early in the morning, the *Hornet* and the *Enterprise* launched 116 aircraft against these targets, including 67 dive bombers and 29 torpedo planes. Because the Americans were still using outdated coordination systems, rather than flying and attacking as one coherent group, the various squadrons became separated and spaced out at different intervals and altitudes.

- Even worse, some of the squadrons followed erroneous courses that would cause them to miss their targets altogether. In addition to these planes, a number of aircraft from Midway had been directed toward the Japanese carriers, including a unit of medium bombers operated by the army, a Marine squadron of dive bombers, and some torpedo bombers.

- When an aircraft carrier is defending itself against aerial attacks, the first and best line of defense is its own fighter planes. Ideally, these intercept the incoming bombers and shoot them down before they can get into range to release their weapons. In combat situations, it was standard procedure among both the Japanese and the Americans to maintain some fighters, known as the Combat Air Patrol (CAP), perpetually circling over their carriers. On the morning of June 4, the Japanese had a strong contingent of fighter planes flying CAP.

- The planes from Midway were the first to reach the Japanese carriers; the CAP shot down 18 of 37, with most of the rest being forced to release their weapons prematurely. Then, a group of 14 B-17s made a high-altitude run, dropping their bombs from where enemy fighters could not reach them. Not a single hit was scored on any Japanese ship.

- Yamamoto had ordered Nagumo to keep half his planes in reserve, loaded with anti-ship weapons, in case the U.S. carriers appeared. Now, Nagumo disobeyed this command and ordered these planes rearmed with bombs to launch a second strike on Midway. Just at this moment, the delayed Number 4 radioed in that it had sighted the American ships.

- Meanwhile, Fletcher had decided to follow the planes that were already in flight from the *Hornet* and the *Enterprise* with those from the *Yorktown*; thus, the final American carrier committed its planes to the battle.

- The first groups of planes from the *Hornet* and *Enterprise* arrived over Nagumo's ships and launched their attacks against the three carriers they could see. Arriving in uncoordinated bunches, the dive bombers and torpedo planes made brave attacks but fell in great numbers to the antiaircraft guns and fighters of Nagumo's fleet.

Nearly 100 American bombers assaulted Nagumo's ships, and still not one hit had been scored.

- But all of these torpedo plane attacks had pulled the protective Japanese CAP fighters down to sea level and scattered them. When several more squadrons of planes from the *Yorktown* and *Enterprise* arrived, the dive bombers were able to attack without interference. In four minutes, three of Japan's front-line carriers had been mortally wounded.

- The final Japanese carrier, *Hiryu*, now launched its planes. Although only a relatively small force, the veteran pilots pressed their attack against the *Yorktown* with determination. Of seven dive bombers, three scored direct hits and two others dropped their bombs close enough to cause damage. The last act of the battle came in the late afternoon, when the Americans sent their remaining planes against the *Hiryu* and set it afire with four bomb hits.

Outcomes

- At the Battle of Midway, the action of just four minutes permanently shifted the balance of power in the Pacific. Before Midway, Japan had a solid edge in numbers of carriers; after Midway, the United States enjoyed a numerical advantage that rapidly and irrevocably increased as the country's industrial capacity came on line.

- Before Midway, the Japanese were always victorious and always on the offensive. After Midway, the roles reversed, and it was the United States that would consistently be on the offensive for the rest of the war and the Japanese who were forced to defend an ever-shrinking empire.

Suggested Reading

Fuchida, et al., *Midway*.

Prange, *Miracle at Midway*.

Symonds, *The Battle of Midway*.

Questions to Consider

1. In what different ways did overconfidence contribute to the Japanese defeat at Midway?

2. Which of the following men was the greatest admiral, and why: Agrippa, Don Juan, Nelson, Yi, Togo, or Yamamoto?

1942 Stalingrad—Hitler's Ambitions Crushed

Lecture 35

In early 1941, Germany was at the height of its power and self-assurance. Its armies and blitzkrieg tactics had proved invincible. Its soldiers were highly trained, experienced, and battle-tested. The German army was at a peak size of 3.5 million men with 3,300 tanks. At the same time, the Russian army was the largest in the world and had as many modern tanks and aircraft as the rest of the world's armies combined. But its men were demoralized, poorly trained, and almost entirely lacking in officers because of Stalin's paranoid purges in the 1930s. Thus, although the Russian army was massive, it had low morale and no leadership. It was at this moment that Hitler would stab his former allies in the back and invade Russia.

Invasion of Russia

- Considering the potential strength of Russia and the size of its army, it seems foolish that Hitler would elect to take on such an enemy and to provoke a war on two fronts, one of the classic errors of strategy. Why, then, did he choose to invade Russia? There are several likely explanations.
  - From Hitler's perspective, by early 1941, it was reasonable to assume that the war in the West was more or less over. The United States had not entered the conflict and did not seem inclined to do so. The only enemy left was Britain, and although the British were safe on their island thanks to their navy, their army posed no threat. The German army was at a peak of training and efficiency, but without use, it would lose this edge.
  - Further, the racist ideology of Hitler and the Nazis led them to believe that the Russians would be easy prey. In Nazi ideology, the Russians were categorized as Slavs, a lesser race of human beings. Hitler assumed that the Russian armies would crumble before his legions of Aryan supermen.

- The code name for the invasion of Russia was Operation Barbarossa, and it was to be a classic blitzkrieg. Armored spearheads would plunge through the Russian lines on narrow fronts, then rapidly speed deep within enemy territory. They would then curve back toward one another, encircling huge chunks of the Russian army. The panzers would carve up the Russians, and the infantry and artillery following the tanks would force the isolated Russian groups to surrender.

- With its army obliterated, it was believed that Russia itself would then surrender. It was never a part of the initial planning to fight a prolonged war or to fight deep within Russia. The Russians unknowingly assisted the Germans in their strategy by deploying most of their army right along the border, thus playing right into the Germans' hands.

- Operation Barbarossa began on June 22, 1941, with a massive artillery barrage and air attacks. Then, the panzers rolled forward. The first few months were ones of unmitigated success and glory for the Germans. The panzers blasted through the Russian lines and cruised deep into Russia. Some panzer groups were averaging 50 miles per day, a rate previously unheard of in war.

- As the invasion stretched into the fall, unease began to creep into the Germans' minds. It seemed that Russia stretched on infinitely and that no matter how many hundreds of miles they advanced, there was always more ahead. And no matter how many men they killed and tanks they wiped out, new ones kept appearing. Even though the Germans were eliminating Russian armies at a phenomenal rate, the Russians were able to replace their losses seemingly infinitely.

- The Germans, on the other hand, could not so easily make good their losses of experienced men or valuable tanks. By the winter of 1941, the Germans had reeled off a string of astonishing victories and had progressed 800 miles into Russia, but they had lost nearly 65 percent of their men and vehicles.

- Finally, the rainy season turned the terrin into a sea of mud, and the German advance began to lose momentum and bog down. Worse was yet to come, however, because the traditional ultimate weapon of Russia was about to arrive—a powerful force that the Russians referred to as "General Winter." By December 1, temperatures plummeted to –35° F.

- Hitler had considered it unnecessary to provide the soldiers with winter clothing because he believed that the Russians would be beaten before winter arrived. As a result, the German troops suffered severe frostbite; the oil in guns jammed, and the guns would not fire; planes could not fly; tank engines froze up; even metal parts became brittle and snapped in the cold. The German advance finally ground to a halt only 40 miles from Moscow, and the Germans settled in for the winter.

The Battle

- By early summer of 1942, the Germans were ready to resume the assault on Russia. Over the winter, Hitler had revised his goals, and oil played a central role. In this era of mechanized warfare, it had quickly become apparent that the machines required huge amounts of fuel to keep running. Germany's only reliable fuel came from Romania, and Hitler was eager to find another source to supply the motors of his planes and tanks.

- Most of Russia's oil came from the Caucasus—80 percent from Baku alone. Obsessed with both the need to obtain new sources of oil for Germany and the idea of cutting Russia off from its own oil, Hitler decided to concentrate his attack on the southern front, with the aim of gaining control over the Russian oil wells in the Caucasus.

- The advances of 1942 would have two objectives: One group would push toward Stalingrad on the Volga River, which contained a number of large factories and sat astride the railways that brought oil from the Caucasus. If the Germans could seize Stalingrad, in theory, they could cut off these supplies. A second German army group would drive even further to the south, directly at the oil fields themselves.

- General Friedrich Paulus was charged with capturing Stalingrad, and his army drove to the city and prepared to assault it. But Stalin decided to make a stand at the city that was named after him. He issued a directive that the Russians would not yield one more step back. To enforce this order, all Russian army units were required to organize detachments of men armed with machine guns who would stand behind the front line with orders to shoot any man who retreated. In August, Stalin put General Zhukov in charge of resisting the German invasion.

In the street-fighting of 1942, the factory areas of Stalingrad were transformed into nightmarish combat zones of rubble, through which small squads of Russian and German soldiers crawled, fought, and died.

- The Germans at first bombarded Stalingrad with planes, reducing it to vast fields of concrete debris. The German soldiers then began moving in, squeezing the Russians into an ever-smaller section of the city. This was a slow process and, in this kind of street-fighting, the German edge in mobility was negated.

- As 1942 wore on, Hitler grew obsessed with the symbolic value of Stalingrad, and he began to divert resources away from the more important drive toward the Caucasus oil fields.
    - Tank units that would have been better employed seizing the oil fields were instead routed to Stalingrad. As Hitler ordered

more troops to be fed into the battle for the city, Stalingrad became the focal point of the entire war in Russia.

- The Russians, too, had sent reinforcements, but they had been playing a clever game in which they committed the bare minimum to the city to keep the Germans occupied while secretly building up a reserve.

- As winter of 1942 approached, the all-important German drive to the oil fields eventually fell short because so many units had been diverted to Stalingrad. Meanwhile, the Russians were down to only a few hundred yards of ground as winter set in, but their objective had been achieved. The Germans had bled themselves dry in the useless battle. With the arrival of winter, the German offensive finally petered out, and now the Russian counterattack began.

- By deliberately not reinforcing Stalingrad, Zhukov and the Russian commanders had amassed a large reserve; with these men, they now attacked the Germans. Two years of war had served to replace the officers killed in the purges of the 1930s with new officers who had experience in modern warfare. They imitated blitzkrieg tactics and turned them back against the Germans.

- Just as he had at Khalkhin Gol, Zhukov used his center to hold the enemy in place while launching pincers to the left and right. Two powerful Russian armies blasted through the German lines on either side of Stalingrad, then trapped Paulus and his entire army. Hitler refused to acknowledge that his attack on Stalingrad had failed and utterly rejected any suggestion that Paulus and his army try to escape while they had the chance.

- By late January 1943, the Germans were no longer able to fight effectively, and Paulus was forced to surrender the miserable remnants of his army. This was the first time that an entire German army had surrendered, and its loss was an enormous and crippling blow to the Germans, both materially and psychologically.

Outcomes

- Although the Germans attempted several more offensives, for the rest of World War II, they were fighting a defensive war. Stalingrad was the turning point of the war against Nazi Germany.

- Before Stalingrad, the Germans retained an aura of invincibility that was utterly destroyed along with Paulus's army. Before Stalingrad, the Germans were constantly advancing and expanding their empire, but after this battle, they were entirely on the defensive, struggling to protect a steadily contracting territory.

- Zhukov and the Russians regained all their lost lands and rolled on into Germany, finally capturing Berlin itself. The Eastern Front ultimately became a war of attrition, one in which the Germans could not compete with Russia's resources. Even though the Germans killed or captured infinitely greater numbers of Russians, the Russians were willing to accept the unequal ratio, and in the end, the Germans were helpless before the vast and unstoppable Red Army.

Suggested Reading

Beevor, *Stalingrad.*

Clark, *Barbarossa.*

Craig, *Enemy at the Gates.*

Questions to Consider

1. What might Germany have done differently to increase its chance of success on the Eastern Front?

2. How much do you think the fact that the city was named "Stalingrad" had to do with both Hitler's determination to conquer it and Stalin's determination to defend it?

Recent & Not-So-Decisive Decisive Battles

Lecture 36

Given the frequency of warfare and its often wide-ranging and dramatic consequences, it comes as no surprise that many individual battles have been decisive turning points. As we've seen, a battle can be decisive for many reasons: a change in rulers, destruction of a state's armed forces and loss of its ability to wage war, the advent of a new technology, or the introduction of social or religious changes. Often, truly decisive battles have more than one of these effects. In this last lecture, we will discuss a few recent battles that might turn out to be among the decisive ones of history and examine some particularly famous battles that may not be as decisive as was once thought.

Arab-Israeli Conflict (1948–1949)

- In November 1947, U.N. Resolution 181 attempted to create two separate states out of the old British protectorate of Palestine—one for Arabs and one for Jews. This sparked ongoing unrest in the disputed territories until, on May 14, 1948, the Jewish state of Israel declared its existence and was immediately recognized by key international powers. The next day, a coalition of Arab states invaded Israel and attempted to destroy it.

- The conflict was fought by a variety of sometimes improvised armies and militias, almost all of them using curious mixtures of whatever leftover equipment from World War II they could lay their hands on. The Israelis thwarted each of the offensives and initiated counterattacks. Between February and July 1949, Israel signed separate armistices with each of the Arab nations involved, bringing to a close the first Arab-Israeli conflict, sometimes referred to as the War for Israeli Independence.

- As a result of this war, Israel ended up with not only the territory given to it by the U.N. resolution but with a good bit of the land allocated to the planned Arab state, as well. Continuing tensions led to the 1967 rematch known as the Six Day War. This war ended with Israel gaining control over more territories, including the Gaza Strip, the Sinai Peninsula, the West Bank, East Jerusalem, and the Golan Heights.

- The outcomes of these wars have set the stage for many of the developments and political crises of the past several decades. Tensions and resentments among Israel, the Palestinians, and various Arab states still constantly threaten to break out into a new conflict. Because of its role in establishing one of the ongoing hot spots of global politics, the Arab-Israeli conflict of 1948 likely deserves a place on the list of decisive battles.

Dien Bien Phu (1954)

- After World War II, France's rule in its old colonial territory of Indochina was challenged by communist Viet Minh insurgents under the leadership of Ho Chi Minh. By the early 1950s, with the Viet Minh receiving support from communist China, the French were struggling to maintain control.

- The French commanders came up with a plan to establish a strong point deep in northeastern Vietnam, which they hoped would disrupt the flow of supplies to the Viet Minh and, perhaps, draw them into a pitched battle, where the technologically superior French troops could destroy them. The site chosen for the French forward base was Dien Bien Phu.

- The Viet Minh responded by moving in 50,000 troops and considerable artillery under the able command of General Vo Nguyen Giap. The result was a seven-week siege, during which the outnumbered French paratroopers were under constant bombardment from hidden Viet Minh artillery and had to fend off numerous attacks by waves of infantry. After a desperate defense, they were at last overcome.

- The defeat at Dien Bien Phu was crushing for the French, both materially and psychologically. Within a few months, the French left Southeast Asia for good, signing the Geneva Accords and dividing the remainder of their former colony along the 17th parallel into independent North and South Vietnams.

- Dien Bien Phu was a decisive battle for several reasons. First, it ended the Indochina War and symbolically closed the door on the era of western colonialism in Asia. Second, it directly led to the Vietnam War, because the United States unsuccessfully stepped in and attempted to oppose further Viet Minh expansion into the newly created South Vietnam. Finally, it set the blueprint for many of the conflicts that followed over the next several decades, in which a major Western power with a high-tech army faced a non-Western, low-tech army that frequently employed guerilla warfare.

The Teutoburg Forest (9 A.D.)

- Among the battles often considered decisive but perhaps not so is the Battle of the Teutoburg Forest, which some have argued was the reason the Romans never conquered Germany as they had Gaul.

- Julius Caesar's conquest of Gaul in the 50s B.C. established the Rhine River as the frontier between Roman territory and the Germania. In the later years of the reign of the emperor Augustus, the Romans began to probe into Germania and form diplomatic ties with some of the seminomadic Germanic tribes.

- The Romans had always pursued a policy of incorporating warlike locals into their own auxiliary forces and granting local elites Roman privileges. Thus, the Roman governor of the region, Quinctilius Varus, had on his staff a young nobleman of the Cherusci tribe who had been given the Roman name Arminius, along with equestrian rank and citizenship.

- But Arminius was only feigning cooperation with the Romans. In 9 A.D., Arminius persuaded Varus to move his three legions to winter quarters by a route that would lead them through dense German forests and swampland. On the march through the forests, the 18,000 men of Varus's command were strung out and vulnerable.

- Arminius's German allies struck the Roman column in a series of hit-and-run ambushes, and a running battle developed that lasted for several days as the desperate legionaries sought to break out of the restrictive terrain while under constant attack. In the end, all three legions were wiped out, making the Battle of the Teutoburg Forest one of the worst defeats in Roman history

- Those who assert that this battle was decisive claim that as a result of it, the Romans never conquered Germany as they had Gaul. Among the cited long-term effects are the preservation of Germanic culture and languages, including its descendant, English. But it is extremely unlikely that the Romans would have conquered Germania or, even if they had, that they would have been successful in imposing Roman civilization.
    - Two hallmarks of Roman civilization were that it was an urban phenomenon that thrived in cities and that it was a Mediterranean culture that prospered best along the shores of that sea.

    - Occupied by nomadic tribes, Germany lacked not only cities, but even any modest-sized towns. It had a completely different climate from the Mediterranean, and the economic system was based on a separate group of crops and foodstuffs. Gaul, by contrast, was already urbanized when Caesar arrived and shared the Mediterranean climate and crops.

    - Conversely, while the Romans might militarily subdue some non-olive-growing regions, those tended to be the same ones that quickly threw off Roman rule or where its culture never displaced the indigenous one. Therefore, although the Battle of the Teutoburg Forest was a dramatic event, it did not determine the cultural future of Central Europe.

D-Day (1944)

- The D-Day landings of June 6, 1944, are often numbered among the decisive battles of history, routinely portrayed as the turning point in World War II—the moment when the fall of Nazi Germany was assured.

- The opening of the second front certainly hastened the end of the war, but a good case can be made that the turning point against the Nazis had happened when the Germans were defeated at Moscow and Stalingrad.

© Getty Images/Photos.com/Thinkstock.

Hitler's Germany was probably already doomed by the time D-Day occurred; although the Allied invasion certainly hastened the end of the war, it was not a decisive turning point.

- It is clear that the Russians would have continued their drive toward Germany, that the Germans would have been unable to stop them, and that the Russians would not have been content with anything less than the destruction and subjugation of Nazi Germany.

- Even after D-Day had thoroughly established the Western Front, Hitler still viewed the Russians as the main threat, and the vast majority of German resources continued to be deployed on the Eastern Front.

- If D-Day was not a decisive factor in defeating the Nazis, however, other actions of the Allies probably were significant. Among these was the annihilation of the German Air Force by the Allied powers. The thousands of aircraft destroyed by the Western Allies might well have made a pivotal difference had they been sent to the east. Another vital contribution to Germany's defeat was the British and American strategic bombing campaign, which greatly reduced Germany's industrial capacity.

- The D-Day landings actually were a pivotal moment in history, although not for the reason usually supposed. Consider what would likely have happened to postwar Europe if the Western Allies had not landed in France and overrun most of Western Europe. There is a good chance that Russian armies would have claimed those countries for the communist sphere of influence.

- In the end, the Normandy landings were a decisive moment in history, not because they saved the freedom of Western Europe from the domination of Nazi Germany, but because they preserved the freedom of Western Europe from domination by Communist Russia.

Conclusions and What Ifs

- Trying to pick and defend a list of the most decisive battles in history forces us to consider the reasons that history has unfolded as it has and to try to discover the complex webs of connections and influences that tie disparate events together.

- Viewing history through the lens of decisive battles naturally leads to a great many questions that begin with the phrase "What if?" Such questions are fun as a kind of mental game, help us comprehend history, and hone our ability to tease out the crucial threads that weave the course of events. Such processes of debate and argument may be the best way both to appreciate the historical process and to gain a deeper understanding of it.

Suggested Reading

Beevor, *D-Day*.

Fall, *Hell in a Very Small Place*.

Glantz and House, *When Titans Clashed*.

Questions to Consider

1. What post–World War II battles do you think should be considered decisive?

2. Which battles would you include on your own list of the decisive battles of world history?

Bibliography

Adkins, Roy. *Trafalgar: The Biography of a Battle*. London: Little Brown, 2004. Popular recounting of the battle.

Amitai-Preiss, Reuven. *Mongols and Mamluks*. Cambridge: Cambridge University Press, 1995. Scholarly book that narrowly focuses just on the fighting between Mongols and Mamluks between 1260 and 1281.

Babur. *The Baburnama: Memoirs of Babur, Prince and Emperor*. W. Thackston, trans. New York: Modern Library, 2002. Entertaining translation of Babur's autobiography.

Beeching, Jack. *The Galleys at Lepanto*. New York: Scribner, 1982. Older standard narrative of the battle and events leading up to it.

Beevor, Anthony. *D-Day: The Battle for Normandy*. New York: Penguin, 2009. Another in the sequence of Beevor's excellent books on the Second World War.

———. *Stalingrad*. New York: Viking, 1998. Well-written and well-researched retelling. The first book I would recommend to read on the epic battle.

Bicheno, Hugh. *Crescent and Cross: The Battle of Lepanto, 1571*. London: Cassel, 2003. Work that focuses more narrowly on the campaign and battle itself than the other books listed.

Bonk, David. *Trenton and Princeton 1776–77*. Oxford: Osprey Publishing, 2009. Typical Osprey publication—short, workmanlike summary, with particularly helpful maps and visuals.

Bradbury, Jim. *The Battle of Hastings*. Stroud, Gloucestershire, UK: The History Press, 2010. Accessible, popular account of William's life and career.

Brett-James, Antony. *Europe against Napoleon: The Leipzig Campaign*. New York: Macmillan, 1970. Extensive collection of primary source accounts covering not just the battle but also the events leading up to it.

Bruce, John. *The Bayeux Tapestry*. London: Bracken Books, 1987. Outdated text, but it contains complete color plates of the entire tapestry.

Bryant, Anthony. *Sekigahara 1600*. Oxford: Osprey Publishing, 1995. Probably the best accessible account in English of the campaign and Battle of Sekigahara.

Cameron, Alan. *The Last Pagans of Rome*. Cambridge: Cambridge University Press, 2011. Up-to-date scholarly study of the 4th century that includes a chapter on the Battle of Frigidus. Cameron argues that the sources have exaggerated the pagan/Christian nature of this battle.

Cannadine, David. *Trafalgar in History: A Battle and Its Afterlife*. New York: Palgrave Macmillan, 2006. Interesting treatment that looks not only at the battle itself but also at what it has come to represent.

Casson, L. *Ships and Seamanship in the Ancient World*. Baltimore: Johns Hopkins University Press, 1995. Scholarly but accessible survey of all types of ancient Mediterranean ships by an authority in the field.

Chasteen, John. *Americanos: Latin America's Struggle for Independence*. Oxford: Oxford University Press, 2008. A good account of the complex story of the independence movements from Chile to Mexico. Much stronger on tracing political events than on military history but helpful for understanding the context of the battles.

Clark, Alan. *Barbarossa: The Russian-German Conflict, 1941–45*. New York: Quill, 1985. Solid overview of the entire story of the Eastern Front in World War II, from the invasion of Russia to the fall of Berlin.

Connolly, Peter. *Greece and Rome at War*. London: Black Cat, 1981. Well-illustrated, influential survey that includes clear descriptions of the weapons and tactics of both the Macedonian phalanx and the Roman manipular legion system.

Coox, Alvin. *Nomonhan: Japan against Russia.* Stanford: Stanford University Press, 1990. Reprint of a massive study (more than 1,200 pages) that heavily relies on Japanese sources.

Corbett, J. S. *Maritime Operations in the Russo-Japanese War, 1904–1905.* Annapolis, MD: Naval Institute Press, 1994. Massive two-volume study of the naval war, with an emphasis on strategic analysis. Rather technical but authoritative.

Cotterrell, Arthur. *Chariot: The Astounding Rise and Fall of the World's First War Machine*. London: Pimlico, 2004. Interesting study of the development and role of the chariot in ancient warfare in the Mediterranean, India, and China.

Craig, Gordon. *The Battle of Königgrätz.* Philadelphia: University of Pennsylvania, 1964. Standard scholarly work on the battle by a famous historian.

Craig, William. *Enemy at the Gates*. New York: Bantam, 1982. Somewhat dated but famous account of Stalingrad, with lots of details drawn from eyewitness accounts.

Creasy, Edward Shepherd. *Fifteen Decisive Battles of the World: From Marathon to Waterloo*. New York: Dorset Press, 1987 (1851). The book that started the modern era of debates over lists of decisive battles. Strong English and Western bias but still worth reading.

Crowly, Roger. *Empires of the Sea: The Siege of Malta, the Battle of Lepanto, and the Contest for the Center of the World.* New York: Random House, 2008. Entertainingly written account of all the conflicts between the Ottoman Turks and Christian Western Europe during the 16th century.

D'Altroy, Terence. *The Incas*. Malden, MA: Blackwell, 2002. Solid and readable survey of Inca history. Strong on economics and politics.

Davis, Paul. *100 Decisive Battles from Ancient Times to the Present*. New York: Oxford University Press, 1999. Well-balanced list of battles, with good, brief descriptions of each.

De Souza, Philip, Waldemar Heckel, and Lloyd Llewellyn-Jones. *The Greeks at War: From Athens to Alexander*. Oxford: Osprey Publishing, 2004. Less on Plataea than on Greek equipment and tactics during this era. A nice introduction to many of the debates about hoplite warfare.

Díaz, Bernal. *The Conquest of New Spain*. New York: Penguin, 1963. Translation of the eyewitness account of the conquest of Mexico by one of the conquistadors who participated. An amazing story.

Diffie, B., and G. Winius. *Foundations of the Portuguese Empire, 1415–1580*. Minneapolis: University of Minnesota Press, 1977. Detailed study of early Portuguese overseas expansion, including accounts of the key battles along the coast of India.

Donner, F. *The Early Islamic Conquests*. Princeton: Princeton University Press, 1981. Standard scholarly account. Stronger on the overall historical narrative than the specifics of the battles.

Dunlop, D. "A New Source of Information on the Battle of Talas or Atlakh." *Ural-Altaische Jahrbuche*r, vol. 36 (1965), pp. 326–330. Very useful article that gives complete translations of the main Arabic sources (al-Athir and al-Dhahabi) for the battle of Talas.

Dwyer, William. *The Day Is Ours!* New York: Viking, 1983. Older but still good account of the battles of Trenton and Princeton that incorporates many nice quotes from primary sources.

Engels, D. *Alexander the Great and the Logistics of the Macedonian Army*. Berkeley: University of California Press, 1978. Exactly what the title claims: an innovative study of how Alexander kept his army supplied and maintained an effective fighting force during his epic march.

Englund, Peter. *The Battle That Shook Europe: Poltava and the Birth of the Russian Empire*. London: Taurus, 2010. Best and most detailed account of the battle in English.

Fall, Bernard. *Hell in a Very Small Place: The Siege of Dien Bien Phu*. New York: Da Capo, 2002 (1967). Detailed account of the battle by a noted war journalist who was later killed in Viet Nam.

Farrokh, K. *Shadows in the Desert: Ancient Persia at War*. Oxford: Osprey Publishing, 2004. Comprehensive analysis of the armies of the Persian, Parthian, and Sassanian empires.

Fischer, David Hackett. *Washington's Crossing*. Oxford: Oxford University Press, 2004. Pulitzer prize–winning book that is very detailed but highly readable.

Fowler, Will. *Santa Anna of Mexico*. Lincoln: University of Nebraska Press, 2007. Biography that is rather sympathetic in its portrayal of the Mexican leader.

Fox, Robin Lane. *Alexander the Great*. London: Penguin, 1973. Solid biography of Alexander that gives a balanced account of his overall career.

Frassanito, William. *Antietam: The Photographic Legacy of America's Bloodiest Day*. New York: Scribner, 1978. Interesting volume of contemporary and near-contemporary photographs of the battlefield.

Fuchida, Misuo, and Masatake Okumiya. *Midway: The Battle That Doomed Japan*. Annapolis, MD: Naval Institute Press, 2001 (1955). Book by a Japanese pilot that is interesting for the insights it offers on Japanese perspectives and experiences regarding the battle, though some of its claims are now disputed.

Gardiner, Alan. *The Kadesh Inscriptions of Ramesses II*. Oxford: The Griffith Institute, 1960. Scholarly translation and commentary on the main sources of information for the Battle of Kadesh.

Glantz, David, and Jonathan House. *When Titans Clashed: How the Red Army Stopped Hitler*. Lawrence, KS: University of Kansas Press, 1995. Important study that draws on previously unavailable Soviet archival material to give a more well-rounded account of the Eastern Front in World War II.

Goedicke, Hans, ed. *Perspectives on the Battle of Kadesh*. Baltimore: Halgo, 1985. A set of scholarly essays about aspects of the Battle of Kadesh.

Goldman, Stuart. *Nomonham, 1939: The Red Army's Victory That Shaped World War II*. Annapolis, MD: Naval Institute Press, 2012. The best up-to-date, easily readable overall account of the battle and its effects.

Gravett, Christopher. *Hastings 1066: The Fall of Saxon England*. Oxford: Osprey Publishing, 2000. Another solid volume in the Osprey campaign series.

Green, Peter. *The Greco-Persian Wars*. Berkeley: University of California Press, 1996. Good overview of the entire course of the wars between Greece and Persia by a noted historian.

Griffith, G. T. "Alexander's Generalship at Gaugamela." *The Journal of Hellenic Studies*, vol. 67 (1947), pp. 77–89. Older but still useful scholarly article analyzing the battle.

Gurval, R. *Actium and Augustus*. Ann Arbor, MI: University of Michigan Press, 1998. Study of the battle, especially as portrayed in Augustan propaganda.

Haldon, J. *Byzantium at War, AD 600–1453*. Oxford: Osprey Publishing, 2002. Readable short survey of Byzantine military organization and tactics.

Haley, James. *Sam Houston*. Norman, OK: University of Oklahoma Press, 2002. An excellent biography, well-researched and highly readable.

Hammond, N. G. L. "The Campaign and Battle of Cynoscephalae in 197 BC." *The Journal of Hellenic Studies*, vol. 108 (1988), pp. 60–82. Scholarly article reconstructing the battle and the events that led up to it.

Hassig, Ross. *Aztec Warfare: Imperial Expansion and Political Control.* Norman, OK: University of Oklahoma Press, 1988. Informative scholarly study of the military system used by the Aztecs to conquer and maintain their empire.

Healy, Mark. *The Warrior Pharaoh: Rameses II and the Battle of Qadesh.* Oxford: Osprey Publishing, 1993. Good basic overview of Ramesses and Kadesh aimed at a general audience.

Heckel, Waldemar, and Ryan Jones. *Macedonian Warrior*. Oxford: Osprey Publishing, 2006. Good, well-illustrated survey on the training of, and especially the equipment used by, the Macedonian phalanx.

Hemming, John. *The Conquest of the Incas*. New York: Harcourt, 1970. Standard narrative that includes a full description of the rebellions and infighting that followed the initial conquest.

Herodotus. *The Landmark Herodotus*. Robert Strassler, ed. New York: Pantheon Books, 2007. Excellent edition of all of Herodotus's *Histories,* with translation, notes, maps, and useful appendices.

———. *Histories Book IX.* Michael Flower and John Marincola, eds. New York: Cambridge University Press, 2002. A scholarly commentary on the main ancient source for Plataea; contains useful information in the notes.

Herwig, Holger. *The Marne: The Opening of World War I and the Battle That Changed the World.* New York: Random House, 2009. Solid coverage of the battle and surrounding events.

Hibbert, Christopher. *Wolfe at Quebec.* New York: The World Publishing Company, 1959. Older, somewhat hagiographic but entertaining biography of Wolfe.

The History of al-Tabari, vols. 11 (Khalid Yahya Blankinship, trans.) and 12 (Yohanan Friedmann, trans.). Albany: State University of New York Press, 1992, 1993. Good translation of the main surviving Arabic account of the battles of Yarmouk and al-Qadisiyyah.

Hofschroer, Peter. *Leipzig 1813*. Oxford: Osprey Publishing, 1993. A somewhat technical treatment of the battle, with a very complete list of all the units involved.

Holmes, Richard. *Battlefield: Decisive Conflicts in History*. New York: Oxford University Press, 2006. Very short blurbs on each battle, in what almost amounts to a survey of the history of warfare.

Joglekar, Jaywant. *Decisive Battles That India Lost*. Lexington: Lulu, 1970. Opinionated but complete description of the first and second battles of Tarain and their protagonists.

Jukes, Geoffrey. *The Russo-Japanese War*. Oxford: Osprey Publishing, 2002. Short but comprehensive and well-illustrated overview of the conflict.

Kedar, B., ed. *The Horns of Hattin: Proceedings of the Second Conference of the Society for the Study of the Crusades and the Latin East*. London: Variorum, 1992. Collection of scholarly essays on the battle and related topics.

Knight, Roger. *The Pursuit of Victory: The Life and Achievement of Horatio Nelson*. New York: Basic Books, 2007. Detailed biography of the famous admiral by a maritime history scholar.

Konstam, Angus. *Poltava 1709*. Westport, CT: Praeger, 2005. Short but well-illustrated narrative of the battle.

Krefft, J. *Ten Battles: Decisive Military Conflicts You May Not Know About, But Should*. Centennial, CO: Military Writers Press, 2009. Contains a good chapter on the Battle of Badr.

Lynch, John. *Simón Bolívar: A Life*. New Haven: Yale University Press. Detailed, even-handed biography of "the Liberator."

Massie, Robert. *Peter the Great*. New York: Random House, 1981. Best-selling and entertaining biography of the Russian reformer that includes a full description of the Poltava campaign.

Mathew, K. *History of the Portuguese Navigation in India*. Delhi: Mittal, 1988. Study of Portuguese colonialism in India, with an emphasis on naval affairs.

Maude, F. N. *The Leipzig Campaign 1813*. East Yorkshire, UK: Leonaur Publishing, 2007. Reprint of a 1908 publication. An old but still useful study of the battle.

May, Timothy. *The Mongol Art of War*. Yardley, PA: Westholme Publishing, 2007. Good modern survey of how the Mongols waged war.

McEwan, Gordon. *The Incas: New Perspectives*. Santa Barbara, CA: ABC-CLIO, 2006. Informative overview of Incan culture and civilization, covering everything from religion to architecture.

McPherson, James. *Crossroads of Freedom: Antietam*. Oxford: Oxford University Press, 2002. Excellent book by a leading Civil War historian that is particularly strong on the long-term influence and importance of the battle.

Millar, Simon. *Vienna 1683*. Oxford: Osprey Publishing, 2008. Short and clear narrative. Particularly strong on illustrations, maps, and the Battle of Kahlenberg that ends the siege.

Moore, Stephen. *Eighteen Minutes: The Battle of San Jacinto and the Texas Independence Campaign*. Dallas: Republic of Texas Press, 2004. Good overall account of the battle and surrounding events that incorporates a lot of interesting quotes from participants.

Morgan, David. *The Mongols*. 2nd ed. New York: Wiley-Blackwell, 2007. Probably still the best general survey of the overall history of the Mongols, by a leading authority.

Morrison, J., J. F. Coates, and N. B. Rankov. *Athenian Trireme: The History and Reconstruction of an Ancient Greek Warship*. Cambridge: Cambridge University Press, 2000. Interesting account of the project to build the *Olympias*, a reconstructed Greek trireme, and explore its abilities.

Nicolle, David. *Yarmuk 636 AD*. Oxford: Osprey Publishing, 1994. Clear overview of the battle and associated campaign, with nice illustrations.

———. *Hattin 1187: Saladin's Greatest Victory*. Oxford: Osprey Publishing, 1993. Solid, short account of the campaign, armies, and personalities involved.

Park, Yune-Hee. *Yi-Sun Shin and His Turtleboat Armada*. Seoul: Shinsaeng Press, 1973. Entertaining but less scholarly account of the admiral's life and achievements.

Peers, C. *Soldiers of the Dragon: Chinese Armies, 1500 BC–AD 1840*. Oxford: Osprey Publishing, 2006. Handy compendium of Osprey titles that covers weapons, tactics, and organization of Chinese armies over a long stretch of time, including the Tang dynasty.

Pleshakov, Constantine. *The Tsar's Last Armada*. New York: Basic Books, 2002. Entertainingly written account of the voyage of Rozhestvensky's fleet and the Battle of Tsushima that is particularly good on quoting Russian sources.

Plutarch, *Life of Flamininus*. Biography of the Roman general by the ancient Greek author.

Pocock, Tom, ed. *Trafalgar: An Eyewitness History*. New York: Penguin, 2005. Useful and enlightening collection of eyewitness accounts of the battle and the key people involved.

Polybius. *History of Rome*. Account of Rome's conquest of the Greek east, including the fullest description of the Battle of Cynoscephalae, by an ancient Greek historian.

Portal, J., ed. *The First Emperor: China's Terracotta Army*. Cambridge: Harvard University Press, 2007. Excellent, well-illustrated collection of articles on Shi Huangdi and his tomb by various scholars.

Prange, Gordon. *Miracle at Midway*. New York: McGraw Hill, 1982. Popular, highly readable retelling of the battle that somewhat exaggerates it as a mismatch.

Reid, S. *Quebec 1759*. Oxford: Osprey Publishing, 2003. Short but solid account with nice maps.

Riley-Smith, J. S. C. *The Crusades: A History*. New Haven: Yale University Press, 2005. Good, accessible, one-volume account of the overall story of the Crusades by a leading historian of the subject.

Robinson, Charles. *The Spanish Invasion of Mexico*. London: Osprey Publishing, 2004. A brief but well-balanced and well-illustrated introduction to the story of the conquest.

Sandhu, G. S. *A Military History of Medieval India*. New Delhi: Vision, 2003. Comprehensive survey of a long span of Indian warfare, covering everything from weapons to logistics.

Sansom, George. *A History of Japan, 1334–1615*. Stanford: Stanford University Press, 1961. Volume 3 of a multipart series on Japan that offers good if somewhat dated coverage of the political situation leading up to Sekigahara, as well as describing the battle.

Sawyer, R. *Ancient Chinese Warfare*. New York: Basic Books, 2011. Detailed look at Chinese warfare that focuses almost exclusively on its earliest periods.

Sears, Stephan. *Landscape Turned Red: The Battle of Antietam*. New York: Mariner Books, 2003. Good, detailed account of the battle, with many eyewitness quotes and nice maps.

Sekunda, Nick, and John Warry. *Alexander the Great*. Oxford: Osprey Publishing, 2004. Well-illustrated book on Alexander that focuses heavily on his battles and his role as a general.

Shaw, Ian, and Daniel Boatright. "Ancient Egyptian Warfare." In *The Ancient World at War: A Global History*, edited by Philip de Souza, pp. 29–45. London: Thames & Hudson, 2008. Up-to-date summary of Egyptian warfare and tactics in a volume that is an excellent introduction to ancient warfare generally.

Shepherd, William. *Plataea 479 BC: The Most Glorious Victory Ever Seen*. Oxford: Osprey Publishing, 2012. Probably the best general introduction to the battle. A solid entry in the Osprey "campaign" series.

Sheppard, S. *Actium 31 BC*. Oxford: Osprey Publishing, 2009. Nice general account of the battle and the campaign leading up to it.

Singh, Harjeet. *Cannons versus Elephants: The Battles of Panipat*. New Delhi: Pentagon, 2011. Detailed look at the three battles of Panipat by a retired Indian Army officer and scholar.

Stacey, C. P. *Quebec 1759: The Siege and the Battle*. Montreal: Robin Brass, 2002. Excellent, well-balanced study of the campaign for Quebec that includes particularly good illustrations, as well as a useful series of appendices of primary source documents.

Steinberg, Jonathan. *Bismarck: A Life*. Oxford: Oxford University Press, 2011. Detailed biography of the statesman that is often critical of its subject.

Stoye, John. *The Siege of Vienna: The Last Great Trial Between Cross and Crescent*. New York: Pegasus Books, 2000 (1964). Classic work on the siege. Getting a bit old but still solid.

Sumner, Ian. *The First Battle of the Marne*. Oxford: Osprey Publishing, 2010. Typical Osprey publication—a short but reliable account of the battle, accompanied by excellent maps.

Symonds, Craig. *The Battle of Midway*. Oxford: Oxford University Press, 2011. Excellent, balanced narrative that takes into account recent scholarship on the campaign.

Thomas, Hugh. *Conquest: Montezuma, Cortes, and the Fall of Old Mexico*. New York: Simon and Schuster, 1993. A readable but very long and highly detailed narrative of the conquest.

Tuchman, Barbara. *The Guns of August*. New York: Macmillan, 1962. Pulitzer prize–winning, highly entertaining narrative of the beginning of World War I, ending with the Battle of the Marne.

Turnbull, Stephen. *The Samurai Invasion of Korea*. Oxford: Osprey Publishing, 2008. Well-illustrated text that is probably the best short overall description of the invasions.

———. *Fighting Ships of the Far East*, vol. 2. Oxford: Osprey Publishing, 2003. Focused study of the structure and design of the vessels used by Korea and Japan during the invasions.

———. *Genghis Khan and the Mongol Conquests*. Oxford: Osprey Publishing, 2003. Typically solid Osprey offering: a short but comprehensive summary with copious illustrations.

———. *Tannenberg 1410: Disaster for the Teutonic Knights*. Oxford: Osprey Publishing, 2003. Short but solid account of the battle, the main figures, and the campaign.

———. *Battles of the Samurai*. London: Arms and Armor, 1987. Book written by a famous military historian of feudal Japan that covers eight famous samurai battles, including Sekigahara.

Urban, William. *The Teutonic Knights: A Military History*. London: Greenhill, 2003. Good, reliable, and readable volume tracing the rise and fall of the order, with an entire chapter on the Battle of Tannenberg.

———. *Tannenberg and After*. Chicago: Lithuanian Research Center, 1999. Authoritative narrative of events leading up to the battle and its consequences. Particularly strong on tracing nuances of politics among the Knights, Poles, Lithuanians, and other groups involved.

Verma, H. N., and A. Verma. *Decisive Battles of India through the Ages*. Campbell, CA: GIP Books, 1994. Brief but clear coverage of the battles of Talas and Tarain and the tactics used.

Wawro, Geoffrey. *The Austro-Prussian War*. Cambridge: Cambridge University Press, 1996. Good account of the entire war that includes a thorough description of the Battle of Königgrätz.

Weir, William. *50 Battles That Changed the World*. Franklin Lakes, NJ: New Page Books, 2004. Offers a somewhat more opinionated list of decisive battles, but the books makes entertaining and thought-provoking reading.

Wheatcroft, Andrew. *The Enemy at the Gate: Habsburgs, Ottomans, and the Battle for Europe*. New York: Basic Books, 2008. Detailed account that includes a description of the campaign leading up to the siege of Vienna and its aftermath.

Wink, André. *Akbar*. Oxford: Oneworld Publications, 2009. Engaging biography of the Mughal emperor by a leading scholar.

Yi, Sun-sin. *War Diary of Admiral Yi Sun-Sin*. Seoul: Yonsei University Press, 1977. Solid English translation of Admiral Yi's interesting diary.

Yupanqui, Titu Cusi. *An Inca Account of the Conquest of Peru*. R. Bauer, trans. Boulder, CO: University Press of Colorado, 2005. Terrific eyewitness version of the conquest from the Inca perspective by a man who was present at most of the key events.

Notes

Notes

Notes

Notes